ОБРАЗОВАНИЕ НЕФТИ И ФОРМИРОВАНИЕ НЕФТЯНЫХ ЗАЛЕЖЕЙ

OBRAZOVANIE NEFTI I FORMIROVANIE NEFTYANYKH ZALEZHEI

ORIGIN OF OIL AND OIL DEPOSITS

ORIGIN OF OIL AND OIL DEPOSITS

by

M. E. Al'tovskii, Z. I. Kuznetsova, and V. M. Shvets

Authorized translation from the Russian

Springer Science+Business Media, LLC

1961

The Russian text
was published by Gostoptekhizdat,
the State Scientific and Technical Publishing House
of the Petroleum and Mineral-Fuel Industry,
in Moscow in 1958

ISBN 978-1-4899-4889-2   ISBN 978-1-4899-4887-8 (eBook)
DOI 10.1007/978-1-4899-4887-8

Library of Congress Catalog Card Number: 60-13948
Copyright 1961 Springer Science+Business Media New York

Originally published by Consultants Bureau Enterprises, Inc. in 1961.
Softcover reprint of the hardcover 1st edition 1961

CONTENTS

Part I

General Questions on the Origin of Oil

Part II

Preliminary Results of Investigations
in the Grozny-Dagestan Oil Region

PREFACE

Attaching great significance in the development of oil to organic substances and microflora contained in subsurface waters, and believing that the presence and quantitative content of these materials may be employed as auxiliary criteria in prospecting for oil and gas, members of the VSEGINGEO Institute (All-Union Scientific-Research Institute for Hydrogeology and Engineering Geology) began in 1953 to study the content of organic substances and microflora in subsurface waters in oil-bearing and non-oil-bearing districts and to investigate the significance of these substances in oil-forming processes.

These studies were made under the direction of M. E. Al'tovskii.

At present, the general theoretical problems concerning the origin of oil have been solved, and investigations are complete on the organic substances and microflora in the subsurface waters in the Grozny-Dagestan oil region.

The authorities of the VSEGINGEO Institute have considered it advisable to publish the results of these investigations in order that the criticism and desires of the readers may more rationally direct the course of all future work. In addition, an acquaintance with this work by a large circle of individuals connected with the study of oil genesis and the prospecting for oil deposits will lead to rapid testing of the hypothesis of oil genesis proposed in this work and will permit an introduction of the useful aspects in actual practice. The given work is considered by the authors to be a beginning of extensive investigations on the composition, conditions of migration, accumulation, and transformation of organic substances in subsurface waters and on the significance of these substances in the oil-forming processes; it is also thought to be a starting point in the study of the geochemistry of the elements closely associated with the chemical composition of subsurface waters.

The work is divided into two parts. The first part gives a critical evaluation of the existing views concerning the origin of oil (chiefly from the hydrogeological point of view), and a new hypothesis is proposed for the formation of the components making up natural petroleum in subsurface waters. The second part presents a brief summary of preliminary results in the study of organic substances and microflora in the subsurface waters of the Grozny-Dagestan region.

Since this book is designed not only for hydrogeologists but also for geologists, chemists, and microbiologists concerned with the study of oil and natural gas, the text has been considerably expanded with data on vertical zoning and on the circulation of subsurface waters and the substances contained in them.

The entire first part of the book and the preliminary notes concerning the status of field investigations in the second part were written by M. E. Al'tovskii; the chapter in the second part concerning the content of organic material in the subsurface waters of the Grozny-Dagestan oil region was written by V. M. Shvets, and the chapter on the results of microbiological investigations in this same region was written by Z. I. Kuznetsova.

The authors will be grateful to readers who will present their critical judgments and who will express their views concerning the future trend that scientific research work should take.

Part I

GENERAL QUESTIONS ON THE ORIGIN OF OIL

INTRODUCTION

The origin of oil and the formation of oil deposits represent one of the most complex problems in geology.

Despite numerous achievements in the geology of oil deposits, in the chemistry, physics, and microbiology of oil, despite many experimental studies on the production of oil-like products from various substances, despite a great number of experiments on present-day migration of water and oil and of gas and oil, and, lastly, despite the abundance of manifold hypotheses that seek, though in the most general outline, to trace oil-forming processes, the problem of the origin of oil remains unsolved.

Judging from the discussions on the origin of petroleum published in the journal "Neftyanoe Khozyaistvo" (The Petroleum Industry) for 1950-1952 and from papers in the collection "Sovetskaya Geologiya" (Soviet Geology) No. 47, 1955, we may state that the various investigators diverge in opinions even in regard to the basic questions concerning the origin of oil and, above all, concerning the initial material. The concept of an inorganic origin of oil persists, even in the middle of the twentieth century, a concept advanced in the seventies and eighties. And, it may appear strange that this hypothesis of inorganic origin of oil is now held, not only by chemists (as seen in the treatise of D. I. Mendeleev), which might seen proper, but also by the geologists N. A. Kudryavtsev and P. N. Kropotkin.

This indicates that discussions on the origin of oil, even at present, are not focused on details or on secondary matters, but on fundamentals, a preliminary grasp of which is necessary to establish a more or less convincing theory that will completely satisfy all our knowledge concerning the geology and chemistry of petroleum.

For all this, however, it is clear that the processes leading to the formation of oil and of oil deposits, processes that are complex and various, demand a joint approach to a solution of the problem, since specialists in several disciplines are required to reach such a solution, particularly geologists, chemists, microbiologists, hydromechanical engineers, and hydrogeologists.

However, this joint approach should involve not only the work of specialists in the various branches of science, pursuing the course previously indicated in our scientific prediction but also a consideration of the problem as a whole from the viewpoint of the laws of each scientific branch, i.e., not only in a secondary approach but in the very basic concept. Only the observance of such a joint effort can lead to highly fruitful results.

As is well known, it is necessary to consider the following factors in attempting to solve the major problem of the origin of oil: a) the initial material, b) the site and form of its accumulation, c) the mechanism of converting the initial material into oil or into its constituent components, d) the accumulation and the conditions of migration of oil, e) the mechanism by which oil deposits are formed, f) the processes of alteration and subsequent destruction of oil deposits.

In considering the origin of petroleum there unavoidably arise supplementary problems relative to the genetic relations between oil and other caustobioliths (for example, natural gas, ozocerite, asphalt, bituminous rocks); and, lastly, there also arises the question of the development of proper criteria to use in prospecting for oil deposits.

The aim of the present book is an examination of the indicated problems from the hydrogeologic point of view, in keeping with the existing data and in agreement with the role of the hydrogeologic conditions, in order to solve any particular problem associated with the origin of petroleum.

In this connection, this book will examine in more or less detail the questions concerning initial material, the site of its accumulation, and the possible alterations of organic substances. Much less attention will be devoted to the mechanism by which oil deposits are formed or to hydrogeologic criteria to be used in prospecting for oil. These matters are included in the book only for the purpose of showing that a different viewpoint concerning the site and conditions of transforming organic substances introduces a fundamental factor in solving problems relative to the formation of oil deposits and to the development of a system of hydrogeologic criteria. This book does not touch on matters, lacking in any single treatment, concerning changes in the quality of the oil during migration and concerning the destruction of oil deposits. This is understandable because subsurface waters, during the course of such processes, exhibit numerous pecularities not related directly to the formation of oil and oil deposits.

Almost all geologists and oil-field workers recognize that subsurface waters are of great importance in the formation of oil deposits. V. P. Baturin [1945], for example, wrote the following: "Horizontal migration of oil, leading eventually to an oil deposit, is governed by currents of subsurface waters. In order to give an immediate prediction of the oil potential, it is not enough to know the paleogeography of the basin, the structure, and the properties of the reservoir rock, it is necessary also to seek to work out the movement of the subsurface waters during the geologic past, to create a new branch of the science — historical hydrogeology."

I. M. Gubkin [1937] attached great importance, in the formation of oil-water emulsions, to the chemical composition of the water mixed with the oil and to the presence of a third phase (suspended particles of fine sand, clay, or calcareous soaps of naphthenic acid). He further maintained that, since the deposition and alteration of organic material occurred in salt water, it was necessary to explain the role of salt water in the transformation of the organic substances. It is regrettable, according to I. M. Gubkin, that this matter has been inadequately discussed. In his paper I. M. Gubkin considered it necessary to present the views of K. Zalozetskii, which are that salt water contains and limits the decomposition of organic substances, halts acid fermentation, and creates an environment of putrefaction, during which the fats are decomposed to fatty acids and to alcohol (glycerin). In the process, the latter are washed away, but the fatty acids are decomposed to hydrocarbons according to the following schemes:

$$C_nH_{2n+1}COOH = C_nH_{2n+2} + CO_2,$$

$$C_nH_{2n-1}COOH = C_nH_{2n} + CO_2,$$

$$C_nH_{2n+2} = C_mH_{2m+2} + C_mH_{2(m-n)} \text{ etc.}$$

There is also no doubt concerning the role of water drive in maintaining particular conditions in the oil horizon during its exploitation. It is well known that one of the present-day methods of most appropriate exploitation of oil deposits is the forcing, by artificial pumping, of water into the oil horizon to maintain and restore pressure conditions in the oil deposit in order to increase the flow of oil to the exploiting well.

There are many views concerning the use of data on the chemical composition of subsurface waters to be used in prospecting for oil deposits, the scientific justification for which may be found only in the assumption that there is some kind of genetic relationship between subsurface waters and oil accumulations and that such waters are not merely some kind of foreign body. It has long been known that highly mineralized salt water has always been considered a favorable sign in prospecting for oil and in evaluating oil potentials in any particular region. However, attempts to relate the presence of oil with the degree of mineralization or with the hydrochemical type of subsurface waters have not proved to be altogether successful, since it has been discovered that actual oil deposits are accompanied by subsurface waters exhibiting several different degrees of mineralization and representing various chemical types. At the same time it has been proposed that, in prospecting for oil, one may use not so much the chemical type of water as the quantity of the various components, a property that is, in the opinion of many investigators, characteristic of particular deposits. Work in this vein has been done by V. A. Sulin, G. M. Sukharev, V. M . Kukanov, D. V. Zhabrev, V. G. Malyshek, and others.

The seepage of oil and water together affects the composition of the combination. In regard to this, V. A. Uspenskii and O. A. Radchenko [1947] wrote: "Seepage, the results of which are clearly manifested in a number of deposits, is extremely widespread, combining with other processes and (with the other processes) leading to changes in the initial type of oil. This phenomenon is apparently very complex in detail, and, despite the fact that it is purely physical, it may introduce profound changes in the composition of the oil. Seepage may be expected to show the effect of adsorption of clay particles, a process clearly expressed in the action of bleaching earths, adsorbing chiefly asphaltenes, tars, and paraffins. Sulfur compounds and naphthenic acids are also adsorbed. Of the hydrocarbons, the aromatic hydrocarbons have the greatest tendency to be adsorbed, the liquid paraffins the least. In general, it may be stated that the light fraction outstrips the heavy fraction of oil during seepage."

But, the very important hydrogeologic factors in the formation, preservation, and destruction of oil deposits have been considered in the works of N. K. Ignatovich, G. M. Sukharev, A. B. Ronov, A. F. Opalev, and N. T. Lindtrop. The basic views of N. K. Ignatovich are shown in Table 1, which is taken from his paper on the appraisal of oil potentials [1945].

TABLE 1

Hydrogeologic Factors in the Stages of Formation, Preservation, and Destruction of Oil Deposits
(according to N. K. Ignatovich)

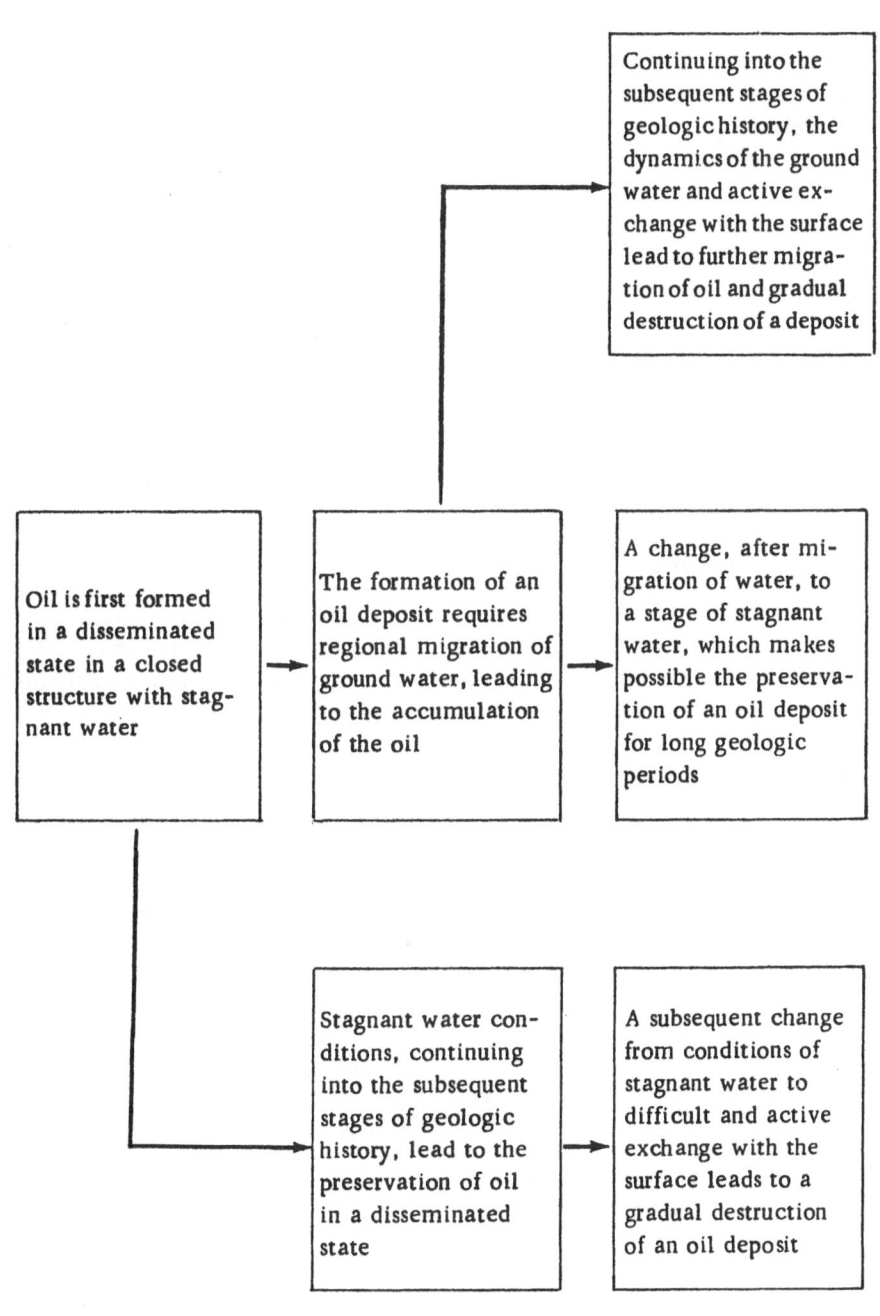

5

Table 1 shows that the principal concept of N. K. Igantovich is that the development of oil-bearing structures and the presence of any particular water conditions in such structures in some cases favor the creation and preservation of oil deposits but, in other cases, lead to the subsequent destruction of the deposits. According to N. K. Ignatovich, not only is a closed structure necessary for the formation of oil, but stagnant water is required as well, and in order for an oil deposit to form, regional migration of subsurface waters is necessary, permitting the accumulation of oil. Without regional migration of underground waters the oil remains in a disseminated state.

And, lastly, a change from stagnant conditions to some stage of structural development in which active water exchange occurs halts the accumulation of oil and even leads to destruction of already formed deposits. In his papers N. K. Ignatovich has maintained that the presence of a structure and the presence of reservoir rocks do not completely guarantee the presence of oil. For example, there is no oil in the Issa structure (Carboniferous in the Penza oblast) or in the Zmieva structure (Permian in Tataria). This fact is confirmed by A. F. Opalev[1947] for a number of structures in Vtoroi Baku. In addition, this author has shown that ground waters having the weakest mineralization (less than $150-200$ g/liter) are characteristic of gas-bearing strata and strata with oxidized petroleum. In general, according to N. K. Ignatovich, zonal distribution of the various ground water types corresponds to zonal distribution of oil and gas deposits.

G. M. Sukharev [1947], who has stated numerous opinions concerning the interrelation of ground water and oil, points out that oil and syngenetic water are present in a disseminated state in initially horizontal strata and that lateral migration through sandy layers to higher levels is necessary for the development of an oil deposit. He has written further the syngenetic water, oil, and gas migrate to the crests during the formation of folds in sandy strata, but that other water is forced toward the limbs of the folds. G. M. Sukharev also cites proof of this. Samples of sandstones beds XIII, XVI, and XXII were collected in the south-eastern segment of a buried anticline in the Oktyabr' region, beyond the limit of oil production. All these samples proved to be saturated with oil, but during testing the sandstone yielded only water, no oil; the same relationship (according to G. M. Sukharev) has been observed in the Staro-Grozny field, in the Tashkala Oisungur, Syuil'-Kort, Makhach-Kala, and Izberbash deposits, and in other localities.

All these statements concerning the role of ground water in the formation, preservation, and destruction of oil deposits and concerning the possibility of using hydrogeologic indicators in evaluating oil potential and in prospecting for oil, as well as the effect of seepage on the composition of oil, undoubtedly correctly emphasize the extraordinary significance of underground water in the origin of oil and the formation of oil deposits. Of course, some of these statements are too generalized and others are insufficiently clear, but they point out the necessity of further investigation in these directions.

However, at present it is necessary to raise, even more widely, the question concerning the significance of ground water in the development of oil and the formation of oil deposits. There are now many data that point to the fact that ground water may be, on the one hand, the geologic agent responsible for accumulating the initial organic material, and, on the other, the medium in which the organic substances are converted into the basic components of oil.

In order to give a somewhat more complete presentation of ground water as the possible medium for oil formation, let us pause for a time on some of the modern concepts concerning vertical zoning and circulation of underground water.

In nature, there are two types of vertical hydrogeologic zones. One of these is represented by zoning that we may observe during well drilling; the other is found along the dips of water-bearing strata. The first should be called vertical zoning of strata, the second vertical formational or, simply, formational zoning. The relationship between these two types of zoning is shown in Fig. 1. This figure shows that only one zone is cut when drilling in the first section (I) — in bicarbonate water; in the second section (II) two zones are cut, and in the third (III), three zones. In examining the change in chemical composition of the water along the dip of the beds, we may always observe a sequential change in the chemical types of water. Since the individual aquifers are almost always separated by impermeable strata, migration of material along the vertical is always more or less difficult.

On the other hand, the movement of water, gas, and accompaning substances downward with the flow of subsurface water, i.e., along the dip of the beds, always occurs with no great difficulty.

It is therefore natural that formational zoning is of most significance in the development of underground water, whereas the zoning that is sometimes observed vertically downward from the surface of the earth is not significant. From here on, in this book, zoning in subsurface water will refer specifically to vertical formational zoning.

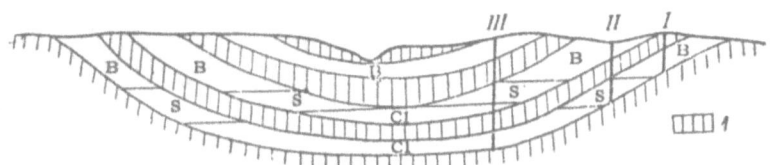

Fig. 1. Diagram showing distribution of formational hydrochemical zones in an artesian system, in which the layers of water have nearly identical rates of movement (the upper horizon is drained). B) Zone of bicarbonate water; S) zone of sulfate water; Cl) zone of chloride water; I, II, III) numbers of sections; 1) impermeable rocks.

The uppermost zone is the zone of soil water, in which the most characteristic processes are the life activities of higher plants and microflora (Fig. 2). Below this occurs a relatively narrow zone of aeration, in which water and air are both characteristically present. Ground water proper is found in the zone of saturation, the upper level of which separates the ground-water zone from the zone of aeration. Oxidation, hydrolysis, solution, and biochemical processes dominate in the ground-water zone and in lower-lying artesian zones in which the water is fresh, with mineral concentrations up to 1-3 g/liter. In deeper zones of artesian, highly mineralized waters (with concentrations up to 300 g/liter and more), salts are precipitated as well as dissolved, oxidizing conditions give way to reducing, and in the upper part of this zone (down to the boundary of physiological activity), biochemical processes are widespread.

Finally, below the last zone mentioned, at depths where the temperature and pressure are above the critical values (above 364-425°C for water), water will occur chiefly in the form of steam, and because of this it tends to rise nearer the earth's surface. As it passes the upper boundary of critical temperature part of this water will condense, forming high-temperature water. The zone should possibly be called the steam-hydrothermal zone, emphasizing the fact that it contains water in the form of steam as well as liquid at high temperature.

It should be added that the first three zones, i.e., the zones of soil, vadose, and ground water, contributing to the development of vertical formational zoning, are subject also to laws of horizontal zoning; for the sake of brevity we shall not dwell on this point here, however.

On the left side of Fig. 2, the closed ellipses and the arrows show the interrelations of the various vertical hydrogeologic zones and the relationship between these zones and the surficial hydrosphere and the atmosphere. The joined right sides of the ellipses for the first five zones underscore the concept that in nature there is a downward flow of water, with contained substances and microorganisms, which consequently migrate from each formation into lower-lying zones. A circle is shown for the steam-hydrothermal zone, indicating, supposedly, that part of the water rising as steam is condensed to the liquid state and again contributes to the supply of water below the critical isotherm, 364-425°C.

The joined left sides of the ellipses indicate upward movements: water in the liquid phase and its accompanying substances through "discharge windows" (papers of A. I. Silin-Bekchurin), water in the gaseous state, and rising streams of various gases, migrating by means of diffusion and effusion. The arrows on the left indicate underground flow in the surficial hydrosphere, but the arrows above indicate supply by meteoric waters and the loss of moisture through evaporation.

Thus, the mutually connected ellipses and arrows indicate a complete cycle of water and its contained substance in the subsurface hydrosphere; they also point out a cycle of water and its contained substance involving the subsurface hydrosphere, the surficial hydrosphere, and the atmosphere.

The above-described hydrogeologic zoning and the related cycles of subsurface waters with their contained substances and microorganisms, as observed by us during the present geologic epoch, may be considered to represent the general conditions for development and migration of subsurface water in the lithosphere and may be thought to be relatively stable for definite intervals of geologic time.

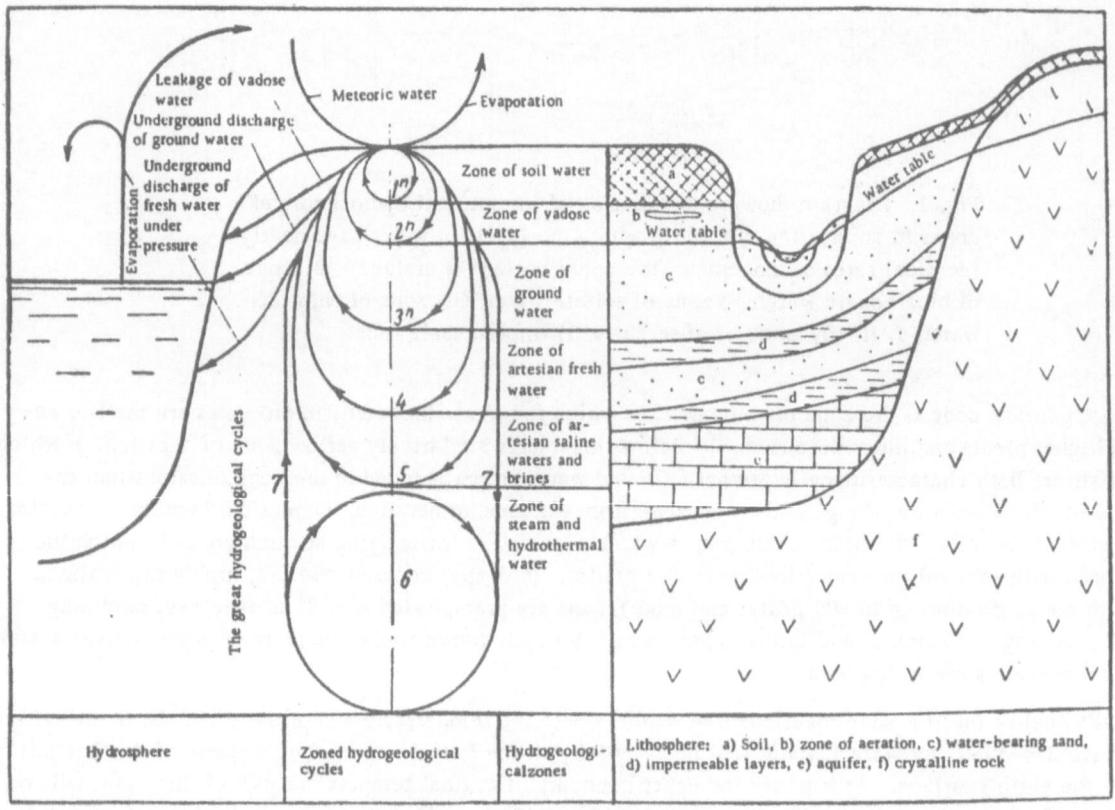

Fig. 2. Diagram of hydrogeological zoning and circulation of ground water and accompanying substances.

This makes it possible to consider the geological development of subsurface water, as a first approximation, to be the development of vertical and horizontal hydrogeologic zoning. This view is in complete agreement with the comparative lithologic method, well known among geologists, according to which the study of modern sedimentation (with certain modifications) is greatly facilitated by an understanding of the formation of fossil facies.

We shall examine briefly any possible changes in hydrogeologic zoning from the historical geology viewpoint.

Ignoring the question of the source of the earth's internal heat, it may still be assumed that the upper surface of the lithosphere was warmer in Precambrian time than in subsequent time. It may further be supposed that volcanic activity was much more intense during the Precambrian than subsequently. The geothermal gradient was very slight. It should hence be assumed that the steam-hydrothermal zone was formed first; then, as drainage patterns developed, the zones of aeration and ground water appeared, and, finally, with increasing growth of the sedimentary layer, artesian fresh water formed and, last of all, waters of high salt concentrations. The entire process of forming new hydrogeologic zones occurred during the simultaneous sinking of the steam-hydrothermal zone into the depths of the lithosphere. The zone of soil water probably did not become completely developed until the end of the Paleozoic, when a permanent plant cover appeared on the continents.

There is no doubt that large-scale changes in the development of the hydrogeologic zones and their regional development have been produced by oscillations of the platform segments of the earth, i.e., alternations of transgression and regression. Transgressions over certain segments of the land have destroyed all the hydro-geologic zones above the level of marine abrasion and have closed off earlier subsurface waters of infiltration origin by a new series of sediments. During regressions, the level of marine sediments attained during the previous transgression is partly destroyed; the sediments are flushed out, thus creating the necessary conditions for a new development of all the upper hydrogeologic zones.

In our opinion, even greater changes in the development of the hydrogeologic zones and their regional distribution are due to folding movements in geosynclinal districts. During differential movements of very large amplitude, when anticlinoria and synclinoria are developed, terminating in the rise of fold-mountain structures above sea level, there is an increase in the amount of heat given off because of the tremendous compression in the rocks and because of the growth of magmatic activity, which draws something of a steam-hydrothermal zone with it to the earth's surface. Hydrostatic pressures give way to hydrodynamic conditions, in which the regions of low pressures correspond to the arches of the folds, especially anticlines that approach the earth's surface. The zones of high pressures occur in the limbs of the folds. Extensive fracturing and even large cavities, up to tens of meters across, may develop in the crests of anticlines; and such openings may be filled during tensional movements with foreign mineral substances introduced from ground waters (according to V. V. Belousov [1954]).

Because of these conditions, all the highly concentrated subsurface waters and brines, tending to move toward the crests of anticlines, are evaporated in considerable measure, leaving behind a series of mineral deposits (common salt, gypsum, sulfur) and many other secondary formations. The mountainous structures that have thus arisen are again filled with subsurface waters of atmospheric origin, and vertical hydrogeological zoning is again established. Thus, fluctuating movements of great magnitude (complete folding according to V. V. Belousov) in geosynclinal regions lead to almost complete destruction of the upper hydrogeologic zones. Such zones are then developed in the uplifted mountain structures, the process commencing anew, but on a different geologic base.

All this leads us to conceive of the development of ground water as a continuous natural historical process, extending throughout all geologic history of the earth and consisting, in turn, of two opposing processes. One of these processes involves the fact that moisture of atmospheric origin tends to fill all pores and fractures in the lithosphere with cold, low-concentration, variable salts (including substances of organic origin) through water that has produced a leached zone down to a certain depth. The second opposing process depends on internal and external heat in the earth, by which waters entering the lithosphere are heated, partly evaporated, concentrated (with the precipitation of various salts, producing a zone of cementation), and, finally, converted through a series of intermediate chemical types to brines having almost maximum salt concentrations.

Tectonic movements destroy the continuity and the smooth functioning of the indicated processes; create new geologic conditions and a new geothermodynamic environment; shift the zones of leaching and cementation in the lithosphere; lead to sharp changes in pressure conditions, to extensive precipitation of salts from ground water, and to separation of gases and other substances; and, in general, shift the process of ground-water development, as it were, to a different stage, at which there begins anew a continuous development until a new phase of tectonic movement occurs.

All these modern concepts concerning hydrogeologic zones and processes turn us afresh to the necessity of profound study of ground water as a basic geologic agent, accounting for fundamental transformations in the lithosphere and undoubtedly being associated with the formation of a number of mineral deposits.

The view was long ago expressed that mineral springs may be considered as mobile mineral veins, since almost identical chemical compounds are gound in both. For example, in the Stimbite Valley, 13 km from Virginia City, Nevada, thermal springs fill adjoining fractures with quartz containing iron and manganese oxides, sulfur compounds of copper, and even traces of gold. I. V. Mushketov [1903], having examined various deposits (travertines) precipitated by mineral waters emerging at the earth's surface, concluded the following: "Conditions favorable for deposition, such as decrease in pressure, temperature, and other factors, are found with equal frequency within the earth's crusts and at springs; spring deposits are therefore widespread, but it is thought that the outer occurrences, which are more accessible to investigation, differ from the internal deposits, which are more complex in composition and structure."

There is now no doubt to us that the substances in ground water (salts, organic material, gases, disseminated elements) may, under favorable conditions, be the source of deposits in the lithosphere, deposits which may occur in quantities of sufficient magnitude to the commercially valuable. It is also clear to us that ground water plays a tremendous role in leaching, transfer, concentration, and deposition of various substances.

It is likely that ground water is of very great importance in the formation of ore deposits and, in particular, of the sulfides of heavy metals.

In general, the study of ground water as a geologic agent should acquire a new significance in modern hydrogeology, since ground water:

a) may be the source (of the salt content) of various salt deposits (rock salt, potash salts, sulfur, and others);

b) may concentrate organic substances and be the medium in which profound changes occur and by which subsequent accumulations of oil deposits and several other bitumens may be effected;

c) may concentrate various disseminated elements, transport them, and deposit them in a new locality and in a new form;

d) is the medium which, in combination with magmatic activity, may produce a number of ore deposits (sulfides of heavy metals and others).

In conclusion, note should be made of the term "oil waters." In this book, in considering oil deposits against a background of systematic circulation of ground waters in the lithosphere and against a background of vertical hydrogeologic zoning, it should be stated that in nature there are no special or individual oil waters. For the most part these are gound waters in the lower zones of water-bearing strata or complexes, being found at maximum stages of concentration.

Therefore, the term "oil waters" in this paper will be used to refer to ground water that is found near or in oil deposits.

Chapter I

SOURCE MATERIAL

The matter of source material is the starting point in seeking a solution to the problem of the origin of oil.

It is precisely at this point that the most basic disagreements are found among the various views concerning the origin of oil. Most investigators, as is well known, think the initial substances were rather stable organic remains. On the other hand, some investigators [D. I. Mendeleev, 1877; V. D. Sokolov, 1899-1913; N. A. Kudryavtsev, 1951-1954; and P. N. Kropotkin, 1955] believe that carbon and hydrogen, which constitute the whole range of hydrocarbon compounds, are of inorganic origin. In favor of the latter point of view a number of more or less convincing facts, taken from nature, may be advanced (for example, the segregation of hydrocarbon compounds during volcanic eruptions, the discovery of such compounds in meteorites, the occurrence of small quantities of oil in igneous rocks); other points include the possibility of producing hydrocarbons by the reaction of water with incandescent metallic carbides and, possibly of most importance, the very convincing criticism of inconsistencies in the views concerning oil-producing or source rocks.

Of course, in speaking of the origin of hydrocarbons in general, and above all of methane, we should recognize that these compounds arise in different ways. During volcanic eruptions, for example, hydrocarbons clearly form by inorganic means. On the other hand, it is equally certain that, in swamps, methane forms by the microbiological decomposition of terrestrial plants.

In this connection, it is extremely important at the very beginning to ascertain what substance or, more precisely, what particular substances we are referring to in any given case when we discuss origins. For all this, there is no doubt that we conceive the problem of oil genesis to signify the origin, under natural conditions, of that complex mixture of substances constituting natural petroleum and accumulating in commercial quantities almost exclusively in sequences of sedimentary rocks. From this it follows that, in studying the origin of petroleum, it is necessary to consider the formation of the gaseous, liquid, and colloidally soluble substances; the hydrocarbon and nonhydrocarbon compounds; the group and individual composition of the hydrocarbon and nonhydrocarbon compounds; and, finally, the principal composition of the microcomponents, generally occurring in a sol of natural oils. In addition, one should always keep in view that we are speaking of extensive regional accumulations of the above-indicated substances in large quantities, but we do include, of course, various small quantities of oil that for some reason or other did not migrate into an oil trap. The total quantity of these substances is known to be considerably greater than the total oil reserves now known on earth.

A very substantial and convincing criticism of the inorganic hypothesis of petroleum origin is to be found in numerous papers of petroleum geologists. In adding to this criticism, for our part, let us note only a few supplementary remarks.

From the hydrogeological point of view, the penetration of surface waters to such great depths in the earth's crust as required by the inorganic hypothesis of petroleum origin is impossible. Such a process is impeded by the pressure of ground water, the value of which exceeds that at the earth's surface in individual regions. The penetration of surface water is impossible also because of the universal development of pressurized artesian strata. In addition, unconsolidated or friable formations cannot long maintain open, gaping fractures, both because the walls are unstable and because the infiltration of water from above always leads to silting and plugging of any fractures. In view of the slight migration of water through any actual natural fractures, it is impossible to conceive the possibility of any considerable quantity of water passing through them (in volumes corresponding to the reserves of petroleum).

If we assume that the earth's atmosphere in its early stages of development was saturated with huge quantities of methane and, possibly, with other simple hydrocarbons, it may then appear possible that such substances were carried into the earth's crust by rain, as gases of atmospheric origin are now introduced into the crust. There are no other actual physical possibilities of transferring gaseous substances from the atmosphere to the lithosphere. But, if this process were the dominant one, the reserves of petroleum in Cambrian deposits should be much greater than in Tertiary rocks, since the supply of hydrocarbons in the atmosphere should have gradually decreased throughout geologic time. But, actually, the reserves of petroleum throughout the geologic systems generally increase, though not uniformly, from the Cambrian to the Neogene.

In general, considering the unconvincing arguments of the inorganic hypothesis of petroleum origin in addition to the numerous and various data indicating that the initial substances were organic compounds, it should be stated that oil is a product of profound reworking of organic substances of plant and, in part, animal origin.

The following data may be offered in favor of an organic origin of oil:

a) restriction of the great bulk of petroleum to sedimentary rocks;

b) oil is but rarely found in igneous rocks and only in small quantities; the presence of oil in such rocks is explained by lateral migration;

c) oil itself is a mixture of various hydrocarbons with admixtures of other inorganic compounds, and its origin is therefore quite naturally associated with organic material;

d) the presence of oxygen, nitrogen, and sulfur compounds in oil, and also the occurrence of other admixtures are most simply and easily explained if the initial substances are assumed to be organic compounds;

e) the presence of porphyrins, which form the basis of chlorophyll and hemaglobin, in oil and several other bituminous rocks definitely indicates an organic origin of oil.

Concepts concerning the initial organic substances of oil have naturally undergone substantial changes in the course of time.

In the first stages of investigating the origin of petroleum, it was supposed that the initial substances were found in particular types of formations (shales, coal beds, peat beds, sapropel) or that the initial substances were the fats of particular kinds of marine animals (herring, shark, whale), or, lastly, of special kinds of marine plants (Zostera and others). In support of this position, more or less ingenious experiments were performed by which various oil-like products were obtained But all these views are now of but historical interest, as indicated by the following.

1. The experiments merely proved that fatty, oil-like products could be extracted from rocks saturated with organic material or from fats, but at no time was anyone successful in obtaining a mixture of the various hydrocarbon, sulfur, nitrogen, and oxygen compounds that constitute natural petroleum.

2. Oil-like products were obtained from the indicated rocks only at high temperatures, a fact in sharp variance with the presence of porphyrins in oils and in several other bitumens; porphyrins cannot exist at such temperatures.

3. The views concerning fats of individual kinds of marine animals and concerning individual marine plants involve a denial of the principle of regionality in the distribution of oil and an affirmation that organic substances accumulate in homogeneous masses, a proposition that also contradicts all known data on the geology of oil deposits and on the distribution of organisms in either continental or marine basins. Some of these views (the hypothesis of K. P. Kalitskii) even deny completely the obvious and demonstrated fact that oil migrates extensively.

Historically this stage of opinion give way to more flexible views, including those in which an entire group of marine or terrestrial plants was supposed to constitute the initial material, or a group of both together; and, later, the stable remains of plants and animals of the whole world were throught to supply the initial substances.

Some investigators proposed that the initial substances consisted of individual components of organic life or of definite products of the decomposition of such life, not having reached complete mineralization.

The investigators ZoBell [1930-1935] and Gibnol [1934] have pointed out that various hydrocarbons form directly in plants, animals, and bacteria as a result of life activity. These hydrocarbons may be preserved in sedimentary rocks and may become the source material for oil. Hecht [1923], Trask [1930], Patnode [1940], and V. P. Baturin [1945] pointed out the possibility of hydrocarbons forming from complex albumin compounds (proteins) as a result of deamidization (loss of NH) and decarboxylation (loss of $-C\underset{OH}{\overset{O}{\lessgtr}}$). Engler [1888], Gefer [1888], and G. L. Stadnikov [1930] assumed that the initial substances came from the fats of planktonic organisms, the activity of anaerobic microorganisms freeing fatty acids, which were further subjected to decarboxylization, hydrogenation (saturation with hydrogen), and polymerization (formation of complex molecules).

V. P. Profir'ev [1930], V. A. Uspenskii, O. R. Radchenko, and Hackford stated that pure carbohydrates may yield oil without fatty substances in an anaerobic environment and in the presence of hydrogen sulfide.

And, lastly, V. A. Sokolov [1925] and N. D. Zelenskii [1936] pointed out the possibility of hydrocarbons forming from methane when acted on by radioactive materials.

Recently, N. T. Shabarova [1953-1954] has expressed the view that the initial substances are organic acids and their salts. According to this view, the organic material of animal organisms is unstable in the presence of hydrolytic and oxidation processes. For example, during mass destruction of fish in fresh-water lakes, the remains of these organisms become undetectable by summer time. As a matter of fact, plant remains, including chlorophyll, accumulate in muddy deposits in lakes. A diagram showing the conversion of organic substances according to N. T. Shabarova is presented in Fig. 3. According to her, the view that oil deposits form from disseminated bitumen is without foundation, because this material is insoluble in water and its content in sedimentary rocks ranges from hundredths to thousandths of one percent, based on the absolute weight of dry remains.

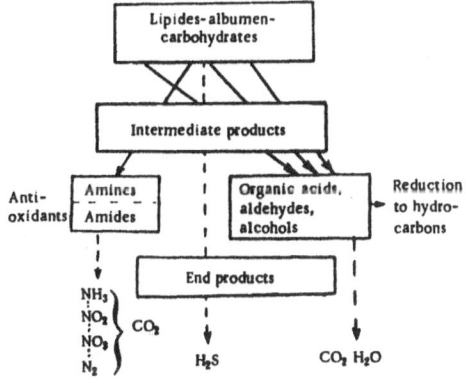

Fig. 3. Diagram showing conversion of organic substances (according to N. T. Shabarova).

However, considering all the present-day information concerning the distribution of organic substances and the variations in composition of the components of natural petroleum, one can hardly maintain that the initial substances from which oil formed belong strictly to any definite group of organic compounds. It makes no difference whether they are albumins, carbohydrates, or organic acids. The problem of initial substances is undoubtedly more complex.

As is well known, all the constituent parts of organic life are completely decomposed when there is an unlimited supply of oxygen, under conditions favorable for the life activity of microorganisms and for the easy removal of the reaction products derived from various chemical and biochemical processes; these constituents are converted finally to mineral substances, water, and carbon dioxide (in an aerobic environment), or to methane and hydrogen sulfide (in an anaerobic environment). For all this, it is clear that these end products can in no manner be converted into that complex mixture of substances which we call natural petroleum. It seems inescapable, therefore, that part of the decomposed organic remnants must be transported from the site of their decomposition into a different environment,[*] in order that the process of disintegration and decay be slowed down, then stopped, and, finally, be reversed, being replaced by a complicated, reducing environment.

From this, of course, one must not conclude that decomposition gives way to reducing processes at one place simultaneously and inevitably for all the substances constituting the organic world.

Let us now turn our attention to the composition of the plant world. On an average plants contain up to 45% C, up to 42% O, up to 7.5% H, up to 1.5% N, and 5% ash that contains various mineral substances. The following substances are found in plants.

1. Cellulose $(C_6H_{10}O_5)_n$, constituting the integument of the cell and representing a very stable substance chemically, being decomposed only by the action of microorganisms;

---

[*]Most investigators believe, quite justifiably, that such conditions are found in an anaerobic, reducing environment.

2. Lignin, which is deposited on cell walls during lignification;

3. Suberin, a fatty substance impregnating cells during the development of cork.

Nourishing reserve substances in plants include nitrogen-free compounds (carbohydrates and fats) and nitrogeneous compounds (albumins). The following substances belong in this group.

1. Starch $(C_6H_{10}O_5)n_2$, occurring in seeds, rhizomes, tubers, stalks, and other parts of plants.

2. Inulin $(C_6H_{10}O_5)n_3$, being held in solution in the cell fluid.

3. Sugar, in the cell fluid of disaccharides $(C_{12}H_{22}O_{11})$.

4. Fats* or vegetable oils, being decomposed into glycerin and fatty acids. They are a mixture of complex esters of trivalent alcohol, glycerin — $C_3H_5(OH)_3$, and higher saturated and unsaturated fatty acids.

5. Organic acids of the cell fluid, monobasic (acetic), dibasic (malic, tartaric, oxalic), and tribasic (citric); intermediate products of exchange are pyruvic and succinic acids.

6. Albumins, arising during the decomposition of a number of amino acids.

7. Cutins, waxy and wax-like substances covering the leaves of plants to decrease evaporation; plants in arid countries have great quantities of cutins (the layer of cutin is 0.5 cm thick on the wax palm).

The principal part of waxes consists of complex esters, formed from high-molecular univalent alcohols and fatty acids. In addition, waxes also contain free fatty acids, alcohols, and hydrocarbons. Waxes occur in small quantities in all plants, but some contain very large amounts. For example, the fan palm yields 7 g of wax from each leaf; a single palm tree may annually give up to 12.5 kg of wax. It should be noted that much wax occurs in brown-coal bitumens. This wax consists of complex esters of high-molecular fatty acids and alcohols (montanic acid and myricyl alcohol). It contains ketone-montonone and, in small quantities, saturated and unsaturated hydrocarbons.

N. G. Titov [1931] discovered 44% nonsaponifying substances in wax from peat; of these, 15% were hydrocarbons, in particular the paraffins $C_{33}H_{68}$ and $C_{35}H_{72}$. The peat was also found to contain a tricyclic hydrocarbon fichtelite, $C_{18}H_{32}$, obtained as a product of decarboxylization and reduction by abietic acid.

8. Oleoresin or balsam, yielding tar, rosin, and turpentine; it occurs in especially large quantities in pine forests, a single tree giving 1-1.5 liters during a season. Oleoresin contains a fluid part — a mixture of terpenes (unsaturated and cyclic paraffins with the general formula $C_{10}H_{16}$) and their oxide derivatives in which resin acids are dissolved.

9. Latex, containing tars, gutta-percha, rubber, sugar, gums, albumin substances, alkaloids, and tannic substances.

The basic substances (protoplasm) contain albumins, which form complex combinations with fatty substances (lipoids) and carbohydrates. In addition, plants also contain pigments of chlorophyll, which is destroyed when the leaves are shed, of the more stable xanthophyll (yellow pigment), which is an unsaturated hydrocarbon $(C_{40}H_{50}O_2)$, and of the chemically similar carotene. The general formulae for the stable substances that may occur in organic remains are shown in Table 2.

During hydrolysis and oxidation, the constituent plant components actually break down, in most cases, to various organic acids. Carbohydrates decomposed to glucose and fructose, from which organic acids are formed. Fats decompose to glycerin and fatty acids; albumins yield various amino acids; wax gives acids and alcohols, and oleoresins break down to terpenes and their oxide compounds.

Therefore, the view of N. T. Shabarova, that organic acids and their salts may be considered the initial substances from which oil is derived, has a very firm basis.

A simple comparison of the above-indicated plant constituents with the group composition of oil immediately shows one that a number of stable components of plants are like several compounds found in oil. The tars

---

*Fats (or lipides), tars, waxes, and hydrocarbons are sometimes joined into a single group, the lipoids.

and waxes may be correlated, in some measure, with the aromatic hydrocarbons, asphalts, and ozocerites; and the nitrogen and oxygen compounds may be compared with corresponding combinations in oil. The salts of a variety of organic acids may be the initial material for the methane, naphthene, and aromatic hydrocarbons.

TABLE 2

General Formulae for the Stable Substances That May Occur in Organic Remains (according to Zaks)

Name	General formula	Particular, best-known individual compounds
Saturated fatty acids	$C_nH_{2n}O_2$	Acids: butyric, valeric, caprylic, lauric, myristic, palmatic, stearic, and others
Unsaturated fatty acids and acids of higher unsaturation	$C_nH_{2n-2}O_2$ $C_nH_{2n-4}O_2$ $C_nH_{2n-6}O_2$ $C_nH_{2n-8}O_2$	Hexadecene, oleic, gadoleic, erucic, linoleic, and linolenic acids
Terpenes	$C_nH_{1.6n}$	Terpenes $C_{10}H_{16}$ and polyterpenes $(C_{10}H_{16})_n$
Acids present in tar	$C_nH_{1.5n}O_5$	Abietic acid $C_{20}H_{30}O_2$
Betulinol	$C_{30}H_{48}(OH)_2$	—
Cholesterol	$C_{27}H_{46}O$	—
Spore-pollen	$C_{90}H_{111}O_{12}(OH)_9$ $C_{90}H_{111}O_{12}(OH)_2$	— —
Lignin	$C_{50}O_{49}H_{11}$ $\begin{matrix}R_1\\R_2\\R_3\end{matrix}$ $C_{46}H_{46}O_{20}$ $\begin{matrix}R_1\\R_2\end{matrix}$ $C_{41}H_{32}O_5$ $\begin{matrix}R_1\\R_2\\R_3\end{matrix}$	— — —
Humic acids	$C_{60}H_{32}O_{24}(COOH)_4 (?)$	—

Natural terpenes (unsaturated cycloparaffins) correspond to the naphthenes or, what is the same thing, to saturated cycloparaffins. And, lastly, the elements in plant ash may correspond chemically with the ash from petroleum. Below we present a number of views of individual investigators, indicating that the stable plant remains may actually be the source material for oil.

G. L. Stadnikov [1937] pointed out that the paraffin character of oil is related to humic substances poor in tars but rich in waxes. In this connection he wrote that the fats of animals are completely destroyed by microorganisms, but that nitrogen and sulfur compounds present in plant organisms prove to be very stable for the most part being capable of persisting with no substantial change through geologic periods and of passing through, without destruction, the incandescent retorts of coking ovens. The stability of these compounds is explained by their cyclic structure.

The above statement refers to sulfur compounds of plant origin; such compounds have been preserved in coal and have thus proved to be stable and to be capable of persisting through geologic periods. The more aromatic hydrocarbons that an oil contains, the richer it is in asphalts and tars, the greater the plant contribution we may attribute to its formation, and, consequently, the greater quantity of stable sulfur and nitrogen compounds there will be in such oil.

According to A. F. Dobryanskii, various tars, balsams, alteration products of terpenes, polyterpenes, and, possibly, lignin and humic acid (i.e., precisely the products of terrestrial plants) generate groups in both the aromatic and the polymethylene series. He also points out that compounds containing phenanthrene are universally distributed in the plant kingdom, but those with anthracene are not; this fact directly relates the complex aromatic hydrocarbons with plants.

A number of laboratory experiments have shown that it is possible to convert plant remains to hydrocarbon compounds. For example, the experiments of Berl [1933] show that, when reducing by hydrogen, carbohydrates are converted to oil-like products, among which paraffin, saturated polymethylene, and aromatic hydrocarbons have been detected. In this connection Berl noted that, when the alkali concentration was high and hydrogenation was effected at high pressures, the mass, representing artificial coal, was converted to a dark brown liquid having a greenish fluorescence, extremely suggestive of oil in its properties.

Laboratory experiments of N. A. Orlov and his coworkers have shown that transformations of acetaldehyde lead to the formation of artificial oil containing chiefly the aromatic series.

In connection with this, N. A. Orlov and V. A. Uspenskii [1934] wrote that laboratory experiments to reproduce natural processes showed that furfural (cyclic aldehyde with a heterocyclic five-member ring) yielded, at the slightest hydrogenation, under pressure, an entire series of hydrocarbons, chiefly of the fatty series, beginning with light pentane and ending with all kinds of lubricating oils having characteristic fluorescence.

All the above-indicated data, of course, cannot be considered as direct and irrefutable proof of the dominant plant origin of oil. But they undoubtedly indicated that, in taking this position, we do not go contrary to the known facts. On the other hand, it is just this position that perhaps best agrees with the modern views concerning destruction of organic substances, concerning the preservation of the stable components, and concerning the presence of particular substances and admixtures in oil, materials extremely important in seeking a solution of the problem of the origin of oil.

Thus, we may conclude that the source material of oil is the group of stable components chiefly of plant organisms. Animal remains are almost completely decomposed and supply a very small quantity of material for the formation of oil, these probably being fatty acids and the coloring matter in blood.

To illustrate the possible contribution of animal remains to the formation of oil, we have cited the data of S. V. Bruevich on the principal products in the Caspian Sea (Table 3).

TABLE 3

Principal Forms of Life in the Caspian Sea (from S. V. Bruevich)

Organisms	Gross wt., million tons	%	Dry wt., million tons	%
Bacteria	2000	61.0	400	75.0
Phytoplankton	1000	30.5	100	18.7
Zooplankton	150	4.5	15	2.8
Zoobenthos	120	3.6	18	3.3
Phytobenthos	3	0.09	0.375	0.07
Fish	3	0.09	0.900	0.16
Totals	3276	~ 100	534.275	~ 100

It is apparent from Table 3 that fish, zoobenthos, and zooplankton represent a small party of the total living assemblage in the Caspian Sea. And, of course, continental animal life contributes even less to the formation of oil. The discovery of porphyrins with an iron complex in oils may indicate that only the coloring matter in blood may be preserved in natural petroleum.

But, beyond this, there arises another very essential question. Which plants, marine or continental, form the source material for petroleum?

It should be emphasized that many investigators hold to the opinion that the source material of oil is found in the stable components of marine flora, since oil forms only in sediments and strata of marine origin.

However, only one, K. Craig [1923], expressed himself on the contribution of terrestrial plants to the origin of petroleum. In this connection, Craig wrote: "The single source material of petroleum, representing at the same time sufficient volume and availability from both the physical and chemical points of view, comes from land plants. . . ." However, Craig state that the remains of land plants accumulate in clays or sands deposited in shallow-water basins and, depending on conditions, may be converted into coal or into oil.

The proponents of this hypothesis offer the following considerations in its favor:

a) the geographic propinquity of oil and coal deposits;

b) transitions from coal-bearing to oil-bearing facies;

c) the production of products similar to oil during the distillation of coal, lignite, and peat.

All these points, of course, rest upon very shaky foundations. Many facts may be marshalled against them, facts that are clearly opposed to the above-listed three statements.

Because of this, the hypothesis of land plants as the source material of oil has been almost completely abandoned. It should be noted, however, that I. M. Gubkin [1937] refers to this hypothesis in a somewhat different manner. He wrote that, if Craig would turn his scientific scrutiny to the other huge segment of plant-animal remains of aqueous origin (i.e., to sapropelitic and sapropelitic-humic deposits), then his theory might lay claim to universal application.

Meanwhile, there is no doubt that substantial migration and accumulation of organic substances in the subsurface hydrosphere force us to turn our attention anew to this point of view, but giving it, of course, an entirely different content and a different basis.

Even at the very beginning of the Paleozoic, the first plants moved out of the water basins onto dry land and quickly covered it. N. S. Naumova has discovered spores of land plants (bryophytes, horsetails, and ferns) in Cambrian deposits of the Leningrad district. According to B. A. Trofimov [1954], the continents had plain, uniform relief during the first half of the Paleozoic, before the Hercynian folding. The plain coasts of the continents, half submerged by water, were favorable sites for the development of swampy stretches and for the growth of terraqueous plants. Warm, moist climates dominated over most of the dry land almost to the end of the Carboniferous. By the end of the Devonian there had arisen on land a great supply of plant food, permitting the abundant development of land animals. The growth of terrestrial vegetation during the Carboniferous increased the quantity of oxygen in the air and facilitated the formation of good soils. In the second half of the Carboniferous and up to the end of the Paleozoic large mountain structures developed (Appalachians, Andes, Urals, Tyan-Shan, Altai), and this led to cooling in the high areas and to the formation of cool, dry zones with temperate continental climate. The land was covered by new vegetation suitable to the changing climatic conditions and consisting chiefly of gymnosperms—conifers, cordaites, and, in warmer localities, ginkgos and ferns. And since oil, according to our view, forms not in muds but in subsurface waters, it is then clear that reservoir rocks (porous rocks) should have formed first and, being depressed to depths in the earth's crust, should have been filled with subsurface water as they sank; this water should then have acquired, at different levels of burial, different hydrochemical compositions, and only then would the conditions necessary for oil formation be created. Thus, land plants of a succeeding geologic period may supply oil to the earlier, such as Silurian plants giving rise to Cambrian oil, and so on. This view agrees with the observations of apparent transitions from coal-bearing facies into oil-bearing facies, and it is in complete accord with the position of L. V. Pustavalov [1941] that epochs of oil formation should follow epochs of coal formation.

In this position, of course, there is no assurance that there is a genetic connection between the coal and oil deposits, but every thought relative to a definite sequence in time of the processes of coal formation and oil formation points to an important genetic relationship, since intensive coal accumulation attests to the luxurious development of terrestrial vegetation and, consequently, to large quantities of organic materials that might have been introduced into ground water during the course of past geologic epochs.

V. P. Baturin has rejected the idea that oil is derived from the remains of land plants only because he found it impossible to point out any fundamentally different conditions on continents for the accumulation of substantial quantities of plant and animal organic material. This is a decisive argument, except for the possibility of very specific liquid hydrocarbons forming on land together with solid biolith coals and sapropels. Actually, as has been shown in this paper, such fundamentally different conditions may be indicated for continental conditions, or, to state it better, for ground waters widely distributed on the continents. Thus, the greatest objection in regard to land plants has been removed.

More profound studies of petroleum chemistry is ever yielding new data indicating that the view that land plants furnished the source material for oil deserves the most serious attention.

For that reason we shall dwell in somewhat greater detail on the contribution that decomposition products of land plants make to the formation of oil. All the more modern views concerning the geochemistry of oil propose, in some degree or other, that humic substances participate in the formation of oil; such materials are thought to be introduced by rivers from the continents to marine basins. But, humus and humic substances are specific products of soil-forming processes, the most characteristic feature of which is the growth and destruction of continental plants.

For example, V. G. Tychinov has separated a nonhydrocarbon substance in the Grozny pipe line, which he reported to have the character of humic acids or of a humic substance. Further, it is very curious to note that E. Holzmann and St. v. Pilat [1931] discovered arachic acid, $C_{20}H_{40}O_2$, in Galician oil; this material is a decomposition product of arachic oil, which occurs in all known peanuts. These peanuts grow at a depth approximately 30-40 cm below the surface. Under such conditions, it is difficult to conceive of the decay products of peanuts being removed by surface streams. It is more likely that such decay products have been washed into ground water by seepage of meteoric water.

It is of interest to turn our attention to the microelement content in petroleum ash, in particular to nickel and vanadium; the first is more abundant in methane, less abundant in petroleum tars. Vanadium, on the other hand, is more abundant in petroleum tars and is almost absent in methane. Both elements, according to the investigations of D. P. Malyug [1933] and of A. P. Vinogradov [1950], are concentrated in the soil cover in great measure by the life processes of land plants. According to I. I. Romm [1946], vanadium is abundant in asphalt-bearing rocks in the Ural region and along the Volga, but it is absent in adjacent limestones and dolomites.

To this we may add that P. V. Smit, quite unexpectedly to him, discovered that the most characteristic microcomponents in the ash from petroleum (nickel and vanadium) are associated not with marine but with continental formations.

Several individual organic compounds in oil also suggest that the source material was derived from the decay products of land plants. For example, pyrodine and quinoline, found in petroleum, are also encountered in coal tar, and this fact points to their origin from continental vegetation.

The presence of various oil-like substances in coal deposits is very interesting from the genetic point of view; these definitely indicate that under certain conditions continental vegetation may furnish the source material for the formation of hydrocarbons in oil.

Thus, for example, K. Craig [1923] reported that bituminous coal was extracted from the Gazi shales in Beluchistan, from the same beds (and the only beds) that yielded petroleum; near Cao de Rolito in Venezuela, flattened tree trunks are observed in bright lignite in Tertiary beds, and all the remainder of the layer yields petroleum.

In Köflach, Styria, oil-like liquid has been observed oozing from lignitic coal. S. V. Kumpan has reported that brown-coal tar has been extracted from sapropelitic coal in the Kuzbas. In 1929, Langecker observed liquid hydrocarbons separating from tarry coal in upper Bavaria (Hodegam).

Recent investigations of carbon isotopes or, more precisely, studies of the percentage variation in the ratio C^{13}/C^{12} also support the view that oil and land plants are genetically related. H. Craig [1953] has shown that the percentage variation from normal is much the same for land plants, modern wood, coal, and oil, the value ranging from -2.4 to -3.0%. On the other hand, for marine invertebrates and marine plants, this value ranges from -1.0 to -2.0%. This isotopic composition of carbon is found both in unrefined petroleum and in the tar obtained after the volatile fraction has been removed. The same isotopic ratios of carbon are found also in the propane and n-butane fractions of oil.

H. Craig, although he points out that the absence of any change in the isotopic carbon ratio relative to the age of limestones, fossil woods, and coal confirms the unchanging value of the ratio from the beginning of the Paleozoic to the present day, nevertheless has written, maintaining the hypothesis of a marine origin of oil as dogma, that the processes of oil formation apparently operate within this ratio exclusively and lead to marked changes in the carbon isotope ratio. However, it would be simpler to suppose that this ratio serves as a direct proof that the source material of petroleum is derived from the stable remains not of marine but of land plants. The view that the source material for oil comes chiefly from the decay products of land plants does not contradict equilibrium considerations or the geologic time of formation of the land plants.

It is known [V. A. Uspenskii and O. A. Radchenko, 1947] that the total mass of coaly-bituminous substance makes up about 1×10^{16} tons, of which 0.001% is found in coal deposits and nearly 0.0003% of the carbon is contained in petroleum. From this it follows that, if a relatively small part of the total mass of terrestrial vegetation might have given rise to the coal deposits, the formation of petroleum deposits would require a considerably smaller part.

Considering all this and assuming a priori that the stable remains of marine flora and fauna must certainly participate in the formation of petroleum, most investigators believe that the source material is derived from the decomposition of both marine and terrestrial organisms.

Therefore, in all the latest papers discussing the geochemistry of petroleum, except for the expressed point of view in regard to its origin, it is invariably explained that rivers carry great quantities of humic material from the land to the sea and that this material is mixed with the remains of marine organsisms. This position is maintained in the papers of I. M. Gubkin [1937], V. A. Sokolov [1948], A. F. Dobryanskii [1948], and G. S. Stadnikov [1937].

In this regard, I. M. Gubkin [1937] has written that not only dead plankton contribute to the collection of material on the floor of a basin; according to him, rivulets, creeks, and rivers carry humic substances to the floor of the basin from the upper part of the soil cover and from eroded peat bogs through which the streams pass in flowing to the basin.

G. L. Stadnikov has emphasized that fats were the parent material of all petroleum. In some cases these fats were formed by great accumulations with no admixutres of other materials of organic origin, but in other cases humic material was added to the accumulation of fats. And he has pointed out, further, that streams carried humic substances containing tars and waxes into basins, to places where vegetable fats were accumulating; in this process the humic materials were mixed with the fats and the entire mixture became homogeneous.

However, another point of view is entirely possible, one assuming that the remains of land plants, with the addition of some decomposition products from continental animals, form the source material and that they accumulate in sedimentary strata by migrating with the ground water.

It should also be noted that many disseminated hydrocarbons form in nature. It is known that they form during volcanic eruptions. They are produced by plants, microorganisms, and some animals, as well as by chemical processes.

An interesting summary of the formation of hydrocarbons during the life activities of organisms has been made by N. B. Vassoevich. Data taken from his paper [1955] are presented in Table 4. With this table N. B. Vassoevich also thought it necessary to cite the following from a paper by B. T. Brooks [1931].

"What is the chemical mechanism of paraffin formation in a rose at the temperature of a June day? It is naturally not thermal decay or distillation pressure or hydrogenation according to Bergius, or exposure of methane to alpha particles, or the result of some kind of mathematical equation."

To this it may be added that normal heptane is found in the "kerosene" nuts of the Philippine plant Pittosporum resiniferine. The stearoptenes of many vegetable oils contain large quantities of the paraffin hydrocarbons. For example, the following hydrocarbons of the methane series have been distinguished: $C_{27}H_{56}$, $C_{31}H_{64}$, $C_{32}H_{66}$, and $C_{50}H_{102}$ from beeswax, Canada wax, the leaves of the palms Coperica and Corypha cerifera, and other plants.

TABLE 4

Hydrocarbons in Living Material

Parts or secretions and kinds of organism in which found	Hydrocarbons
	Paraffin Series
Rose	Sclid paraffins
Many species of eucalyptus, Compositae, Umbelliferae, etc.	Solid paraffins
Leaves of the Ouricuri palm	Normal paraffins from $C_{24}H_{50}$ to $C_{36}H_{74}$
Several species of Pinus (conifers)	Heptane (liquid) C_7H_{16}
Wax from tobacco leaves	Heptacosane $C_{27}H_{56}$
Wax from apple peel	Heptacosane and nonacosane $C_{29}H_{60}$
Wax from the leaves and cork tissue of many plants	n-Pentatriacontane $C_{35}H_{72}$ (melts at 75°C)
Candelilla, nearly 5% by weight of the dry plant consists of wax	Hentriacontane (40-60% wax) $C_{31}H_{64}$
Liver of marine vertebrates	Isooctadecane $C_{18}H_{38}$
	Terpenes with open chain
Essential oil, secreted by many plants	Myrcene (liquid) $C_{10}H_{16}$
Basil oil	Ocimene (liquid) $C_{10}H_{16}$
Shark-liver oil	Squalene $C_{30}H_{50}$
Milky sap from land plants (rubber is found in more than 2000 species of plant)	Caoutchouc $(C_5H_8)_n - C_nH_{2n-2}$
Wild rosemary	Myrcene (or its isomer) $C_{10}H_{16}$
	Cyclic terpenes (liquids, more widespread in nature)
Sap and pitch of conifers and other plants	Pinene $C_{10}H_{16}$
Oil from conifer needles, lemon oil, etc.	Bisabolene $C_{15}H_{24}$
Lemon oil, orange peel, celery and caraway oil	d-Limonene $C_{10}H_{16}$
Essential oils of plants (celery, etc.), tars in eucalyptus oil	Celinene $C_{15}H_{24}$
Essential oils of pine needles	l-Limonene $C_{10}H_{16}$
Vegetable oils of many plants	Cadinene $C_{15}H_{24}$
Sequoia (needles)	Sequoiene $C_{13}H_{10}$
	Other hydrocarbons
Vegetable pigment of tomatoes, carrots, etc., and also some bacteria	Carotene $C_{40}H_{56}$
Several bacteria	Leprotene and other caratinoids
	Hydrocarbons (undifferentiated)
Diatoms	Hydrocarbons
Penicillium	Liquid hydrocarbons
Serratia marinorubrum, Vibrio ponticus, and other marine microbes	Liquid and solid hydrocarbons, optically active
Sulfate-reducing bacteria	Methane and naphthene hydrocarbons
Oysters	Naphthene-methane hydrocarbons
Marine algae	Hydrocarbons

The formation of the hydrocarbon mineral fichtelite from abietic acid bears witness to the formation of disseminated hydrocarbons. Thus, we should apparently not seek the mechanism of hydrocarbon formation from the geologic point of view, since the entire process may possibly reduce to the accumulation of already present but disseminated hydrocarbons.

It should be noted in conclusion that the above-indicated three points of view (the accumulation of marine and terrestrial plant remains in marine basins, the accumulation of land plant remains in subsurface waters, and the accumulation of disseminated hydrocarbons) are not, in essence, mutually exclusive. It is possible that they supplement each other, and this fact leads us to recognize the need of more profound study to appraise the specific significance of the sources of initial substances for the development of petroluem: organic decay products, the conditions of their decomposition, migration, transformation, and accumulation. In this study we should consider not only the burial of the material in marine sediments but aslo the extent of its migration through the atmosphere, the surface and subsurface hydrosphere, and the lithosphere, where it finally accumulates in oil and gas deposits.

Chapter II

SOURCE BEDS

The consensus of opinion, if we exclude the proponents of an inorganic origin of petroleum, is that the initial organic substances accumulate on the floors of marine basins, where they are mixed with various mineral assemblages (chiefly clay particles) and where they form sedimentary masses, subsequently being transformed into varieties of bituminous rocks. In this way source rocks or oil-producing strata are formed.

These views are based on the apparent, but completely unsupported, geologic rule that marine muds, saturated to a more or less degree with organic material, actually form most of the known bituminous rocks.

Such views are strengthened by the universally known fact that an overwhelming proportion of industrial oil production comes from sedimentary rocks of marine origin.

But, did the sediments actually form in marine basins, and did these sediments later become rocks possessing extraordinary, peculiar properties that led to the production of oil?

This question must be considered an entirely appropriate one, despite the fact that the views on source beds for petroleum are the basis of modern hypotheses concerning the origin of petroleum, such views being more or less universally accepted.

Ideas concerning source beds arose long ago. As early as the middle of the last century G. Abikh [1867] wrote concerning the migration of oil from deeply buried rocks rich in organic material. Views more closely resembling modern opinions were advanced by N. I. Andrusov and G. P. Mikhailovskii toward the end of the last century and at the beginning of the present century. But these ideas, depsite their almost century-long existence, have not solved the problem of the origin of oil. Even at the present time, there are no real discussions on the physicochemical and biochemical processes or on the stimuli leading to the conversion of the stable organic remnants into oil; nor are there any explanations of the physicomechanical processes or stimuli regulating the so-called initial stages of the joint migration of liquid oil, water, and gases,. i.e., the movement of these constituents into the reservoir rocks. Even views on the lithology and diagnostic features of the source beds are extremely indefinite. Over the span of almost half a century there has been practically no development in the position that organisms are buried in muds, where they are converted to oil, and that the oil is then squeezed out by pressures of the overlying strata into reservoir rocks. This alone furnishes grounds for a serious, critical review of opinions concerning source beds.

A number of leading petroleum geologists deny, in general, the existence of source beds in nature. For example, K. P. Kalitskii has indicated [1911] that clays occurring between oil-bearing rocks do not contain oil; this phenomenon has been observed in outcrops and in samples from wells. According to K. P. Kalitskii, if oil can be extracted only with difficulty from sands (washed by acid or forced by gas or water), how can clays easily yield or transmit oil. He has written further that when people speak of source beds, which are generally thought to be clays, they find it possible to maintain that these clays yield up their oil completely. During migration oil should, as dye liquids do, leave traces everywhere, in the form of stains and solid petroleum products. When oil migrates along joints, the pressure difference between source locality and accumulation site cannot be assumed to be constantly maintained. The pressures must eventually equalize and movement must cease. Consequently, each oil deposit should have a root extending out from it to the source bed. Further, K. P. Kalitskii, from experience in drilling practice on oil deposits, pointed out that oil is confined to sands and is entirely absent in clay rocks that separate one oil horizon from another. Another leading petroleum geologist, N. A. Kudryatsev, believes that views concerning the burial of organisms in muds and the discovery of oil in sedimentary rocks are hypothetical. In his paper [1951], he makes numerous convincing criticisms of petroleum source beds.

It is still of interest to note a communication of I. O. Brod [1954]. He wrote that until now it has been impossible to prove that oil-producing organic material is related exclusively to the oceanic envelope, or, in contrast to this, to find definite distributions of oil and gas accumulations throughout the great extent and throughout the numerous horizons of Paleozoic epicontinental deposits of the Russian and North American platforms and throughout almost the entire extent of the intermontane basins of Central Asia, which are filled with rocks of Mesozoic age (thought to be continental sediments).

By analyzing the above-mentioned considerations one cannot and must not agree with the hypothesis of K. P. Kalitskii or with any hypothesis of an inorganic origin of oil; the extraordinarily slight yield of liquids from clays under natural conditions, the negligible permeability of shales, and the laws of pressure distribution between two reservoirs (source and reservoir) and the fractures connecting them remain irrefutable physical phenomena and laws, and we must inevitably come around to consider them.

Therefore, the discussions of many years, relative to views on source beds, have stirred some investigators to make more detailed studies on recent and fossil muds and formations that have been considered source beds.

The first investigation of this type was begun by A. D. Arkhangel'skii in the Grozny region [1926]. Noting all the imperfections in the existing methods of determining carbon in rocks and muds, A. D. Arkhangel'skii pointed out that the Chokrak-Spirialis and the Spaniodontella sands differ in no essential way from similar formations of any age or any other deposit. The organic material they contain (carbon) they acquired from clays or from the carbonized plant remains in the clays and from veinlets of carbonaceous material, all of which are present in very small quantities (from 0.1 to 1%, i.e., constituting approximately from 2.7 to 27 kg per cubic meter of rock). There are clays in oil-bearing horizons with greater contents of organic material. Clays in the Novo-Grozny region contain from 56 to 58 kg carbon per cubic meter of rock, and in the Staro-Grozny region the carbon ranges from 44 to 77 kg per cubic meter. Bituminous limestones contain up to 14.7% carbon. In this connection, a carbon content of 40-45 kg per cubic meter of clay rock is grounds for considering the rock a source bed of petroleum. But, in the same paper, A. D. Arkhangel'skii noted that Moscovian Jurassic clays contain up to 5.5% carbon, and the Jurassic oil shales in the Volga and Obshchii Syrt region have as much as 20-30%. There is, however, no oil deposit of any kind associated with these rocks. Extensive investigations on the organic content and its group composition in recent and fossil muds (i.e., on supposed source beds) have been made by P. D. Trask. He has analyzed 30,000 samples of soils and muds. It would seem that such a tremendous volume of work should, if not solve the problem of the origin of oil in source beds, at least establish diagnostic features of such series. However, this has not come to pass.

Recently, a number of other papers have appeared in connection with the study of organic substances in recent muds. For example, N. T. Shabarova, as a result of such investigations, points out [1953] that organic material in recent marine sediments consists of lipides (bitumens), nitrogeneous substances of an albumin nature, and nitrogen-free compounds of polysaccharide types. In addition, there are nonhydrolyzed remains. In her opinion, complete decomposition of organic material occurs on the floor of the basin (italics by M. Al'tovskii). A single intermediate decay product is represented by organic acids, chiefly of the fatty series, which react with inorganic compounds dissolved in water and form salts of the organic acids; some of these reactions products remain in solution, some precipitate in the sediment. Thus, according to N. T. Shabarova, neutral compounds are present in the sediments, i.e., aldehydes, alcohols, fusel oil, and salts of organic acids (italics of M. Al'tovskii).

According to the investigations of V. G. Savich [1950], recent marine sediments contain organic acids in quantities shown in Table 5.

Organic acids detected include acetic, butyric, and formic.

Interesting investigations have been carried out by M. I. Subbota [1952] on the definite content of the average bitumen "A" in near-shore and offshore muds in the Caspian Sea near the Apsheron Peninsula. She wrote that the content of bitumen in near-shore sandy muds ranges from 0.00125 to 0.0025% by weight, the quantity of bitumen decreasing in samples taken at greater depths. For example, the bitumen content in samples taken from point No. 71 at a depth of 1 cm proved to be 0.01%; in the interval from 0.4 to 1.0 m, samples contained only 0.00125%; and in the interval from 1.0 to 1.4 m, the amount decreased to 0.00031%. A similar picture is seen in offshore muds. As a result of these investigations, M. I. Subbota concluded that organic substances are converted, to a considerable extent, to bitumens on the floor of the marine basin, in the very uppermost part of the basin. Another part of the organic material, being more stable, remains in an unaltered form and may be

TABLE 5

Content of Organic Acids (Volatized with Steam) in Recent Marine Sediments

Type of sediment	Amt. of organic acids, mg per g of sediment	pH	Eh, mv
Organic muds, forming by decomposition of algae (lagoon of the Black Sea)	10.54	7.6	− 214
Organic muds, the organic part forming by decomposition of flowering plants (coastal zone of the Caspian Sea)	3.227	8.28	− 91
Black coquina sand	0.761	7.56	− 142
Brown silt	0.654	7.8	+ 5
Dark green clayey silt	0.658	7.8	− 1.0

preserved in this state for an indefinite time, depending on the reigning geochemical environment during diagenesis of the sediments. Methane, according to the investigations of M. I. Subbota, is found as traces in muds. Apparently most of the methane escapes as a free gas, moving upward out of the sediments, but the rest is destroyed by methane-oxidizing bacteria.

According to investigations of V. V. Veber and N. T. Shabarova [1953], the accumulation of organic material in the sediments of lagoons and semiclosed marine basins is chiefly due to the decomposition of bottom layers of various benthonic plants.

These observations indicate that, in general, the group composition of organic substances (in percentage ratios), beginning with the initial material, gradually evens out, although some individual pecularities in the investigated species of plants are preserved.

In their works, V. A. Uspenskii and O. A. Radchenko have devoted considerable attention to bitumens in sedimentary rocks. In their last work [1954], they have differentiated three ways in which the mineral part of sedimentary rocks incorporates organic material.

1. Adsorbed organic material, invariably affixed to mineral particles and uniformly distributed throughout the rock, in contrast to oil bitumens; adsorbed organic material contains up to 20% hydrocarbons; the hydrocarbons may appear by synthesis in the bodies of microbes or they may result from bacteria or ferments decarboxylizing organic acids, i.e., in approximately the same way as the minerals fichtelite and scheererite form from abietic acid in peat.

2. Organic material as cement in the rock.

3. Organic material in detrital form.

In summary, these authors have written that the results of their investigations indicate consistently that in nature there do not exist different regional forms of primary occurrences of hydrocarbons in sedimentary rocks, except for negligible quantities in syngenetic bituminous material. In addition, they have expressed doubt that the amount of hydrocarbons can possibly endow a rock with the character of a source bed. Syngenetic bituminous material is taken to mean adsorbed organic material.

The presence of bitumens of organic carbon (substances extracted from the muds by organic solvents), fatty acids, humic substances, hydrocarbons, and even cerotic acids (constituting the principal part of beeswax) in modern muds and partly indurated sediments has been established by the investigations of P. D. Trask, K. F. Rodinova, V. V. Veber, A. N. Gorskaya, E. A. Glebovskaya, P. V. Smit, and several others.

The investigations of A. N. Gorskaya and E. A. Glebovskaya have shown that the hydrocarbons in Caspian Sea sediments are chiefly methane and naphthene. Practically identical data were obtained by P. V. Smit in his studies of muds in the Gulf of Mexico.

Investigations in the Northern Caucasus during an expedition of the All-Union Scientific-Research Institute for Geological Exploration uncovered the presence of disseminated hydrocarbons in the Tertiary clays in the Northern Caucasus in amounts of 0.003-0.01%, or approximately from 0.2 to 0.8 kg per cubic meter of clay. According to V. S. Muromtseva, these hydrocarbons also belong to the methane-naphthene group.

According to V. A. Uspenskii [1955], who has studied the organic material in rocks in great detail, it has been demonstrated that during coal formation, in the broadest sense of the word, only a small part of the initial organic material, by virtue of peculiarities in its nature, is converted to mobile products that later, being driven into a reservoir rock, may give rise to the accumulation of oil. This author believes that the principal part of the initial organic material gives rise to coal-like products and that it constitutes the ordinary, universally observed component of organic material in rocks.

From this it follows that the great bulk of organic material buried during sedimentation and diagenesis is not of the bitumen (hydrocarbon) series but of a coal series, which came from the first by alteration.

The most important result of all these investigations is the discovery that small quantities of disseminated hydrocarbons, chiefly of the methane-naphthene series, are present both in modern and in partly indurated muds, and also in clays that have already passed through all the stages of diagenesis.

The most likely source of these hydrocarbons is found in the hydrocarbons that form in the bodies of bacteria. These hydrocarbons are extracted from rocks by preliminary mechanical comminution and subsequent treatment with various organic solvents.

These same investigations have furnished rather convincing proof that the process of hydrocarbon formation in nature is exceptionally developed only in the oceanic zones of the earth and in the sediments deposited in these zones.

In this connection, we should recall some data concerning the formation of hydrocarbons in various natural environments. Hydrocarbons do not form only in the bodies of marine microorganisms; they also form in terrestrial organisms, as well as in microflora that are widespread in ground water.

Many of the higher plants on continents also form paraffin hydrocarbons and terpenes with open and cyclic chains (see the data in Table 4).

Hydrocarbons from organic remains are also formed, as is well known, in various natural environments. Thus, according to studies of P. V. Smit, hydrocarbons are found in deposits of deltas, swamps, muddy shallow lakes, rivers, loess, soils, and peat. This investigator has also noted that sandy marine sediments contain approximately eight times the quantity of hydrocarbons that silty clays in contact with these sands contain. Discovery of the minerals balkhashite (elaterite) and coorongite indicates that hydrocarbons may form in an aqueous environment even under aerobic and oxidizing conditions. Interesting occurrences are observed in the naphthoids, tiny inclusions of oil-like liquid imprisoned generally in some of the volcanic rocks of Georgia, Bulgaria, the Scandinavian Peninsula, South and North America, Manchuria, and other localities. Of no less interest are ozocerite-like minerals such as chrismatine in Saxony and urpethite in England and Hungary. These oil-like minerals have a semiliquid consistency. They are confined to coal deposits.

The hydrocarbon minerals fichtelite and scheererite form in peat deposits from abietic acid.

In general, with increasing thoroughness in techniques of determining organic substances, organic carbon, organic acids, bitumens, and hydrocarbons in natural occurrences, it becomes more and more clear that these materials may form in widespread regions and under a variety of conditions. Therefore, an objective evaluation of data obtained on the discoveries of hydrocarbons in nature requires that we make more profound studies on these substances in surface basins at contacts on the floors of the basins, in muds and rocks, and in underground water from the atmosphere, in soils and ground water.

In this connection, the presence of hydrocarbons in modern and partly indurated muds cannot yet be considered complete proof that these muds are the true sites of oil formation. The proof, as stated by mathematicians, is necessary but not sufficient. Actually, the presence of carbonaceous substances and carbon minerals in oil-bearing rocks or the presence of liquid bitumens in coal deposits is not yet proof that oil is produced from coal or, on the other hand, that coal is formed from oil; nor can the presence of hydrocarbons in volcanic eruptions be considered absolute proof of the inorganic origin of all.

At present, the most we may do is state firmly that natural petroleum does not form in the mud stage of sediments, since, if this were true, traces of oil formation and of natural oil itself should be found in modern muds as frequently as, for instance, peat is encountered on the continents. Hydrocarbons occur in muds; nonhydrocarbon compounds of organic origin are also present; but as yet natural petroleum has not been discovered.

If the processes of oil formation belong to the later stages of diagenesis, this practically eliminates completely the effect of such potent factors as the life activity of bacteria, for which, as is true for all living organisms, life is impossible without water (or out of water). To propose that the processes of compaction, dehydration cementation, and recrystallization of sediments (leading to the formation of rocks) favor the development of microflora or processes of oil formation is, of course, to state something that has no real foundation. To be precise in this discussion it should be stated that when one writes of "modern muds," referring to materials having a moisture content of 40-60 and even up to 80%, he is not considering real mud, i.e., more or less compacted sediment, but suspended mineral particles in water that lies on the actual floor of the aqueous basin.

In fact, many investigations have confirmed the fact that microbiological activity dies out almost completely at a very shallow depth beneath the floor of the aqueous basin. For example, at a depth of 21.6-21.9 m below the floor of the Gulf of Mexico, 4000 aerobic or facultatively anaerobic bacteria and 2200 anaerobic bacteria were found in one gram of sediment. At a depth of 138.3-138.5 m, the numbers of bacteria declined to 100 and 20 respectively.

Lastly, if it is proposed that the final products of decay are organic acids and their derivatives, adsorbed by the mineral components of the sediment, it then becomes necessary to consider the state of the particles according to the laws of colloidal physical chemistry. The adsorbed organic material may cover the mineral particles by virtue of molecular attraction; ions of organic acids may occur in a diffused layer of micelles or be present in a hydrosol. Not one of these states or environmental conditions possesses the stimuli or is able to create a mechanism for converting organic substances to oil. Mechanically adsorbed material may remain in this condition without change for an indefinitely long time; acid ions in a diffused layer of micelles may begin processes of adsorption exchange (and which may reach completion instantly), and hydrosols may be converted to gels in the course of time, solidifying and either forming part of the cement of the rock or contributing detrital fragments to the rock.

In short, if by diagenesis we imply, not something indefinite, but chiefly the processes of compaction, dehydration, cementation, recrystallization, and other processes of lithification, it becomes difficult to conceive a more favorable environment for the extensive development of a group of simultaneous and sequential complex reactions, which may reduce almost completely the mineralized remains of organic substances to a very complex mixture of liquid and gaseous hydrocarbons.

The hypothesis of source beds, representing a very complex process of transforming the remains of organic substances, shifts the scene from an active aqueous environment to a physicochemically slightly active environment, at first in mud-sediments and then in rocks. Actually, each constituent part of a marine sediment represents a final product of some kind of process. The mineral composition is the final product of decomposition of rocks; inorganic compounds are salts precipitated from a marine solution after saturation for that ingredient has been reached; gases from all reactions in the sea escape upward and are not precipitated in the sediment. Therefore, the physicochemical activity of sediment is incomparably lower than the marine solution from which it has precipitated. And only in the mud stage, i.e., when the sediment is to a considerable degree saturated with water, can intense biochemical processes favorable to the life activity of microflora be effective; as soon as the mud becomes compacted and the water is squeezed from it, and this occurs at very shallow depths, biochemical processes will begin to slow down considerably and, after all the water is squeezed from the sediments, will cease or will come to completion only in pores filled with water.

Increasing the temperature of sediments by burial can scarcely activate the sediments physicochemically, since experiments on distillation, beginning with the experiments of Engler, have shown that this process requires rather high temperature (above 200-300°); and these cannot exist during processes of oil formation. All these data lead us to assume that the conversion of initial organic substances to oil does not take place in the inert sediments but in the physicochemically and biochemically active environment of an aqueous solution.

No satisfactory hypothesis of source beds can be found at this time in any mechanism, from the point of view of modern concepts concerning soil mechanics or hydrodynamics of liquids and gases, during the initial stages of migration of the components of petroleum from the source beds to a reservoir. Let us examine this problem in somewhat greater detail. Let us note at the very beginning that, for water and oil, being liquids, it is necessary to keep in mind only hydrostatic pressure if the layers containing them are in any degree hydraulically connected. If there is no hydraulic connection between beds or if the connection is restricted, each stratum will have its own pressure and the movement of liquids (migration) can occur only when there is a difference in these pressures.

We might further speak of pressure of the mineral framework of the beds, due to the individual constituent particles (grains, platelets of clay minerals, crystals, and other solid ingredients). But this pressure has no relation to the fluids (water and oil) migrating through cavities in the rock if the framework has already acquired a definite form and stability under certain geologic conditions.

It is known, for example, that marine muds (data from S. V. Bruevich) contain up to 60-80% and more water; i.e., their apparent porosity is 60-80%. The porosity of most compact clays in their natural state, including marine clays, is about 20-30%. When such muds are compacted under the pressure of some engineering structure (buildings, dams), this porosity is somewhat decreased; the structure settles slightly and the process then becomes stabilized. Loosely distributed sand grains have a porosity of 48%; in a more compact distribution the porosity becomes 26%. Massive, crystalline rocks have porosity of a different kind, amounting to 1-2% and less. The porosity of clays at great depths (on the order of a thousand meters or more) may decrease to only a few percent.

From this it follows most clearly that in nature, except for the zone where rocks exist in a plastic state, there are no strata without porosity, and, consequently, water and oil cannot be completely squeezed out of the rocks containing them. Not only should traces of oil remain, but, if the porosity of clay persists, as it generally does, at 20-30%, rather large quantities should be left in the source beds. However, this phenomenon is not observed in nature, not on a regional scale at least.

Experiments on the compaction of sands and other rocks have shown that oil is not squeezed completely out of them even at pressures of 7030 kg/cm^2 and that the bulk of the liquid (that is removed) is squeezed out in the first hour and a half, even within the first 15 minutes. In this process the squeezed-out material may only migrate into rocks whose pores are filled with compressible material, such as gas, not into rocks saturated with incompressible water.

V. D. Lomtadze [1951] performed some interesting experiments on this subject. He mixed water and oil with clay until the mass was capable of flowing. Clays used in the experiments were Cambrian silty clay, kaolin (consisting of 95% kaolinite), bentonite (containing 70% montmorillonite), and marshallite (consisting of sandy, quartzose, slightly argillaceous dust). The mixture was placed in a steel cylinder, and, by compressing the cylinder head, the pressure drop (difference between pressure in the cylinder and atmospheric pressure) was gradually increased to 1000 kg/cm^2. In this way, it was found that the porosity of the Cambrian clays decreased only to 11%, of the kaolinite to 10%, of the marshallite to 15%, and of the bentonite only to 37%.

Experiments by N. A. Buneev, P. A. Kryukov, and E. B. Rengarten [1946] of using steel cylinders to squeeze water out of Jurassic clays and Upper Carboniferous limestones (the specimens having their natural water content) yielded the following results: only 64.6% of the moisture in the pores of the Jurassic clay could be squeezed out when the sample had a natural moisture content of 31% and the unilateral pressure drop ranged from 1450 to 6000 kg/cm^2; the Upper Carboniferous limestones, with a natural moisture content of 9.8%, began to yield water only at a pressure drop to 4300 kg/cm^2, and, at 6500 kg/cm^2, 50% of the natural moisture was driven out.

The experiments of S. L. Zaks [1947-1952] on squeezing water and oil simultaneously from argillaceous and sandy rocks led him to the conclusion expressed in the following sentence. If there is little oil (10% of the pore volume), and water, relative to the oil, is abundant, only water, not oil, is extracted.

When the oil content reaches 35%, water and oil begin to come out together, but later, water alone. When water and oil occur in equal quantities (or oil is more abundant) pure oil is driven off at first, then oil alternating with water, and afterward pure water.

All these experiments have shown that tremendous drops in pressure, on the order of several thousand atmospheres per square centimeter (pressures that do not occur under natural conditions) are necessary in order for only a part (not all) of the contained water or oil to be squeezed out of sandy clay rocks. An oil is squeezed out only if it is present in volumes approximately equal to one-third the volume of water. In addition, we must always keep in mind a very important circumstance that, regardless of whether the oil is driven from the source beds by hydrostatic pressure or by pressure transmitted through the framework of the rock, muds and clays are compressed only by differences in pressure and not by absolute pressure, such as that which may exist at any particular depth below the surface of the earth. From this it follows that if the sands capable of accumulating oil occur at depth, say 1000 m, and an underlying clay lies at an average of 1050 m, compression occurs at a pressure of only 5 kg/cm^2, i.e., only approximately wtice the load exerted by engineering structures. Under these conditions nothing can be said about whether the clay is more or less completely compacted. Clays under such conditions will preserve a very high porosity. This is everywhere observed in nature. Pressure differences under natural conditions have such small values that shales preserve even the morphological elements in plant imprints (as seen in Silurian shales, for example). High and even very high pressures differences, from the viewpoint of compression may be found only during times of tectonic movements, but if we agree with the view that compression is attained only during tectonic movements, oil should be preserved in clays for many millions of years.

Everyone knows that any rock may yield as much water as it is capable of absorbing. This phenomenon is used to determine the permeability of rocks to water by forced pumping, to restore pressure conditions in oil deposits by flooding, and to effect drainage of higher aquifers by draining into lower-lying aquifers. All computed formulae for pumping water are good for forced flooding if the sign is changed from minus to plus. But in practice, as everyone knows, clays and clay rocks absorb water very slightly or not at all. Consequently, they possess this same capacity in relation to the squeezing out of water or any other kind of liquid, including oil. But, in regard to oil, clays and clay rocks should be considered impermeable with all the resulting consequences in relation to both primary and succeeding migration of oil. Thus, it may be said if clays and clay rocks had not acted as impermeable barriers to oil and water there would not have developed individual water-bearing horizons (aquifers), there would be no separate subsurface lakes, seas, and oceans, and there would be no oil and gas deposits, since these latter form only when the layers bounding them are very slightly permeable to oil and gas.

Thus, V. A. Tabasaranskii [1954], in analyzing the conditions under which oil and gas deposits formed in the Il'skii-Kholmskaya region, has indicated that a study of deep-hole cores in that region, in the interval below the productive horizons, shows no traces of oil; this suggests that a vertical migration of hydrocarbons out of the Mesozoic rocks took place, even though samples abound in numerous microfractures and larger joints. In his opinion, only lateral migration took place in this region.

All this appears the more valid when it is realized that the movement of water and oil is accomplished not under the influence of hydrostatic pressure existing at some particular depth, but because of pressure differences, which in nature, according to a tremendous quantity of hydrogeological data, are very small. The ground-water gradient is approximately 0.0001. Such a gradient secures only a very slow movement of ground water along a bed, and even thin beds of clay under such conditions stop vertical movement almost completely.

In this regard, the investigations of V. N. Shchelkachev and M. A. Gussein-Zade [1953] are of interest; these were made to discover the possible volume of water discharging upward or downward during pumping to flood the Tuimazy field. The principal conclusion made by these investigators is that, when the permeability of the impermeable beds between horizons (Devonian clays and silts — M. Al'tovskii) is measured in thousandths or ten thousandths of a millidarcy, if the impermeable coffer beds are a few meters (or more) thick and if they are not fractured, it is impossible to force any substantial amount of liquid from one bed through the intervening impermeable layer that separates it from the overlying or underlying horizon.

Some of the results of these studies by V. N Shchelkavech and M. A. Gussein-Zade are presented in Table 6.

It should be further noted that, almost universally, the relations of pressure in subsurface water of water-bearing horizons is such that each upper bed has a greater pressure than a lower-lying formation; i.e., in most cases only migration upward from below is possible, not the reverse. Such a relation in the distribution of hydrostatic pressure in the zone of sedimentary strata may be substantially changed only during tectonic movements.

Recently, in connection with experimental investigations of M. A. Kapelyushnikov and his coworkers [1952], the reciprocal processes of evaporation and condensation have been proposed as the principal means of transfer of petroleum from source beds to reservoir rocks. In considering this it is necessary to recall the conditions of the experiments and their principal results. Oil and several gases in various proportions were placed in a container in which a constant temperature of about 70° was maintained during the experiment and into which the gases were introduced, under a pressure of several hundred atmospheres. In addition, the investigators tested the effect of the reciprocal evaporation of the solid phase and of the solid phase together with water. This was accomplished by placing divided cartridges in the vessel in which powders of dry or moist rocks were loaded.

TABLE 6

Results of Computations by V. N. Shchelkavech and M. A. Gussein-Zade on the Flooding of the Tuimazy Field

Coefficient of permeability of the roof rock, millidarcies	Pressure in the upper horizon, atm	Distance, m	$\dfrac{Q_y}{Q_H}$
0.00001	100	5000	0.0009
0.0001	100	5000	0.0088
0.001	100	5000	0.083

Explanation: k, the coefficient of permeability for the horizon is 700 millidarcies; t, the thickness of the horizon is 20 m; b, the thickness of the clay roof is 10 m; R_n, the pressure for flooding is 180 atm; R_e, the pressure in the producing wells is 100 atm; Q_y is the volume of water discharged; Q_n is the volume of water pumped down for flooding.

An increase in the thickness of the clay roof rock (in this case taken as 10 m) should show a proportionate decrease in the coefficient of permeability of the layer.

The principal results of these experiments, interesting to us in this present problem, may be summarized in the following.

1. For the processes of reciprocal evaporation to be effective a very high pressure drop, relative to natural conditions, is necessary: from 100 to 500 atmospheres and more.

2. In order to change oil to a gaseous phase it is necessary that the volume of free-gas mixture be several times the volume of oil (3-5 times).

3. The critical pressure is lower the smaller the specific gravity of the oil; heavy, high-sulfur oil with a specific gravity of 0.9077 cannot be converted to gas even at pressures above 500 atmospheres.

4. The presence of water in the rock raises the critical pressure 10-55 atmospheres.

Furthermore, the authors have noted that the first part of methane introduced into the container causes a separation of asphaltic-tar compounds from the oil, the material settling on the walls of the vessel. The authors consider the results of their experiments on tarry oils to be "merely approximate."

The phenomena of reciprocal evaporation and reciprocal condensation, of course, have not been doubted by anyone. Nor is there any doubt that these phenomena find their chief significance in the formation of so-called gas-condensation deposits. But, whether these processes are of wide regional extent and whether they represent the mechanism whereby oil is transferred from source beds to reservoir rocks, that is entirely a different matter. The following remarks may be made on this question:

a) pressure drops on the order of hundreds of atmospheres may occur in nature only during tectonic movements, in beds being folded;

b) the constituent components of oil scarcely form in this way in order for free gases to accumulate in volume several times that of the liquid phase; the gases are first found in solution, not in a free state;

c) if the water in the powder placed in a cartridge increases the critical pressure 3-15%, it is necessary to make clear how much this pressure will increase if the entire container is filled with a mixture of clay and water;

d) the processes of reciprocal evaporation when methane has been introduced cannot be explained by the migration of heavy, tarry, and high-sulfur oils, since this phenomenon produces fractionation of the oil, as a result of which the asphalt-tar compounds remain in place.

But, the chief difficulty, perhaps, lies in visualizing the real, physical relationship between the laboratory, thermostatically controlled vessel and a source bed. If the latter is conceived of as a single vessel in which reciprocal evaporation occurs, what is the source of the necessary mixture of gases (introduced under great pressures), since the free gases separating within the source bed before the liquid phase will tend to be disseminated in the direction of lower pressures? If the source bed with its huge dimensions is thought of as a system of small isolated, thermostatically controlled vessels, the isolated centers effectively accomplish a physical migration of the gas phase, to say nothing of the fact that each such small vessel should be considered a focus of independent and vigorous generation of free gases, activated by unknown causes.

In general, we may hold to the opinion of O. A. Radchenko [1951] in this matter; in regard to the processes of reciprocal evaporation he wrote the following: "However, the very scale of the phenomena and the restriction of deposits of the indicated kind" (petroleum distillates also — M. Al'tovskii) "to regions where they are oil deposits of the ordinary type lower in the section prevent us from adopting this scheme as a universal explanation."

Thus, the hypothesis of source beds until now has presented to mechanism for transferring the petroleum from the source bed to the reservoir rock, however, plausible it may seem from the viewpoint of modern knowledge on soil mechanics, hydrodynamics, and the physics of the aggregate state, or from the viewpoint of actual existing conditions in the lithosphere continuously filled with ground water.

The hypothesis of source beds automatically leads to a completely improbable concept of the migration of oil from the hydrodynamic point of view. Actually, clays should be impermeable to oil, as they are to all other liquids, whereas porous and jointed rocks should transmit readily; open faults in crystalline and massive rocks may also be good sites for migration, but in unconsolidated or friable rocks faults make underground barriers; movement of oil should occur only when there is a pressure difference, not as the result of absolute hydrostatic pressures, and the velocity of flow should depend on the length of the path and the fall (or the piezometric slope). In a number of papers the views on possible means of oil migration from the source beds are clearly unacceptable; they contradict the basic laws of hydrodynamics, a matter V. P. Savchenko turned his attention to in his discussion of 1950-1952.

There are still some proper remarks to make concerning the impossibility of finding suitable source beds at many oil deposits, however similar they may appear to oil-producing strata; i.e., it is impossible to point out some bed or beds with a high content of organic material which might be, however complex the path of migration, the source of the petroleum. Deposits of this type include those in the basins of Central Asia; they occur in continental Mesozoic deposits, which rest on strongly metamorphosed Paleozoic rocks and on Jurassic volcanic rocks. Similar deposits are found in America and China. If we assume hydrodynamically improbable migration of oil, we may, with great difficulty, find some possibly suitable source beds for deposits in Western Turkmenia, the fore-Ural depression,* Georgia (the Shirak series), Ukhta, and even the Apsheron Peninsula. If, when seeking a solution to the problems of oil migration, we are guided strictly by the laws of hydrodynamics, we are obliged to seek for a proper source bed for each individual oil deposit, and we should find it directly at the contact with the reservoir beds. But, as a consequence of this, the number of oil deposits having no formations at all resembling source beds becomes very large.

In general it may be stated that we have no direct or indirect data permitting us to propose the real existence in nature of oil-producing or source beds, as there are also no sufficient grounds for assuming that the initial organic material accumulated only in marine basins.

However, the accumulation of huge quantities of organic material in marine basins in an undoubted geologic fact. But, such accumulations correspond, not to oil, but to various bituminous rocks, in which the organic material represents syngenetic mineral constituents in greater or lesser stages of decomposition and transformation. The carbon in oil constitutes only the 0.000006 part of the total quantity of disseminated organic material in the lithosphere (this is six ten-thousandths of one percent). Therefore, from the viewpoint of balance, the tremendous accumulation of organic material in marine basins represents almost 99.994% of all organic material that has accumulated in bituminous rocks. Oil shales are formed when the content of organic material is high in clay rocks.

---

*The presence of small quantities (several tons) of semiliquid, gassy bitumens in the Domanik series of Tataria and Bashkiria proves nothing, since this series is composed not only of mudstones but also of limestones and marls that are known to be highly permeable. The consideration of the Domanik series as source beds for petroleum has been disputed by K. B. Ashirov [1954].

V. A. Uspenskii and O. A. Radchenko [1954] wrote the over-all conversion of buried organic material occurs, in fact, according to a single general plan, by carbonization, i.e., by a process that represents the beginning of coals and oil shales and also the huge masses of disseminated organic material of the same nature as the organic material in the coals and oil shales. These investigators believe that oil, in this complex multistaged process, is merely a very subordinate (in mass) secondary product.

A fact frequently cited in support of the view that the source material for petroleum accumulates in marine basins is that most oil deposits and most economic reserves of oil are found in strata of marine origin. This fact is true. But, this does not furnish proof that the oil formed in marine sediments. Actually, oil is found not only in marine strata but also in friable continental deposits, even in those of Quaternary age,* as well as in various volcanic rocks. It is well known that there are many examples of such accumulations of oil. Some explain the presence of oil in volcanic rocks and in continental deposits by assuming the most improbable conditions and means of migration of the oil. It is most acceptable, of course, to assume that the distribution of oil is in no way related to the origin of the rocks or to their ages but that it is a function, basically, only of the physical properties of the oil-bearing rocks. Therefore, the restriction of oil deposits to marine rocks does not represent a law, there being several exceptions; but it does represent a phenomenon that should lead us to conclude that marine beds possess the best reservoir properties and that they are regionally widespread.

In general it seems to us that the hypothesis of source beds is supported only by the most precarious foundation and that we have every reason to try a different way and to try by other means to solve the fundamental questions concerning the origin of oil, concerning the place the initial organic substances accumulated, and concerning the place where this material was converted to oil.

---

*The Novobogatinskoe region (Emba) and the Caspian terraces (Dagonin).

Chapter III

PROCESSES OF OIL FORMATION IN SUBSURFACE WATER

In the Introduction to this book it was pointed out that, in seeking a solution to the problem relating to the origin of oil, one means of migration of the organic substances has been neglected and completely uninvestigated. This course is through the atmosphere, soil, and ground water, passing into the deepest, buried segments of the lower structures of the lithospheres, where oil and gas deposits are chiefly concentrated. As will be shown, this course of migration and manner of conversion of organic substances undoubtedly exist in nature and are important factors in the formation of oil and gas. To say that anything definite in this respect has been firmly established is, of course, very difficult. The answer to this question requires extensive investigation, based on the most recent methods of studying organic substances and microflora in natural underground solutions. The task will naturally require the collective efforts of men in various specialties. Publication of the material presented below will serve, it seems to us, to arouse interest in the investigations of these specialists, and, even more, it will fulfill one of the important tasks facing the given investigations.

It is known that the atmosphere contains aerial organic compounds, including many hydrocarbons. A tremendous number of these compounds participate in various soil processes. Some of these substances undoubtedly drop out of the soil cycle and during years, centuries, many thousands and millions of years, contribute to the essential composition of the ground water.

Further, organic substances that fall into the ground water move with the water, and, over a very long period of time, in any case on the order of ten thousand or several hundred thousand years, they pass through a series of physicochemical environments (alkaline, oxidation zone of weakly mineralized water; alkaline, reducing, sometimes sulfurous zone of highly mineralized water; and, lastly, alkaline, reducing zone of brines that contain various microcomponents) with gradual increase in temperature, approximately from 7 to 100-150°C, and gradual increase in pressure, from atmospheric to several hundred atmospheres.

The principal source of organic material for this course of migration is naturally terrestrial vegetation, which, being decomposed to stable components, surely begins to fall with rainwater into the soil cover, almost everywhere and, during further seepage, it migrates into the ground water.

In this chapter we shall attempt, so far as is possible with the existing data, to lay a foundation for, and then develop, the above-stated position.

First of all, let us pause to make several very preliminary remarks.

I. M. Gubkin has noted a very interesting feature in the geographic distribution of oil deposits. This feature is that the overwhelming majority of deposits occur in zones bordering mountain chains, at sites where the mountain structure may be buried and secondary, considerably weakened folding may be developed, and in great geosynclinal zones between large mountain regions.

From the hydrogeological point of view, this pattern of distribution, as observed by I. M. Gubkin, signifies that most oil deposits are found in systems of pressurized water or, as stated in hydrogeology, in artesian basins, which contain highly mineralized water and brine in their deepest parts. If, geologically, the above-indicated pattern of distribution signifies that oil deposits are closely associated with tectonics, hydrogeologically, this relationship means, above all, that not only oil deposits but the oil itself is intimately associated paragenetically with pressurized water in artesian basins. And, since artesian systems develop most commonly in downwarped layers of marine origin, there is nothing surprising in the fact that the greatest oil fields are associated with marine sediments.

However, highly mineralized artesian water may be found in continental formations (as in several basins in Central Asia, for example), and, consequently, oil deposits may also form in continental beds. In some, but considerably less common, cases, artesian water occurs in volcanic and metamorphic rocks, and it is possible that these too may contain oil deposits. From this, it follows that the genetic relations between oil deposits and ground water are closer than between oil and tectonics. On the other hand, there is no clear genetic relationship between oil and the origin of the rocks, since oil and oil deposits, as everyone knows, may occur in rocks of the most diverse origins.

In establishing a genetic relationship between oil and highly mineralized artesian water we are demonstrating nothing particularly new, since the presence of salt water, encountered when drilling for oil, has long been considered a favorable indication. Here we wish only to emphasize and examine more closely these already known genetic relationships.

In the petroleum literature there is a widespread view that water, oil, and gas are confined to petroleum reservoir rocks, i.e., to rocks possessing considerable porosity (intergranular or fracture). From the viewpoint of a restricted zone, i.e., where there is an oil deposit and an oil-water interface, this may be true, if we are careful to keep in mind that a single rock is at once the reservoir rock for oil, gas, and water. But, if we keep in mind that the water at the outer margin of the oil deposit or, more widely, any water associated with the oil deposit represents but a small part of the water-bearing strata forming the artesian system, it then follows that oil fields or oil (and gas) deposits may be said to be confined to aquifers which, in comparison with the oil or gas deposits themselves, have a very wide regional distribution.

Oil (and gas) deposits, relative to the water-bearing horizon they occur in, are very small, local accumulations of those particular substances constituting natural petroleum. A close connection has been established even between oil deposits and the conditions obtaining in supplying the water-bearing horizons in which the deposits occur. For example, Herold [1941] noted that the general productivity of the Kern River field on the eastern slope of the San Joaquin Valley in California increases when the winters have been moist for a number of years; a number of dry years leads to the opposite effect. Oil and gas deposits may therefore be compared, in this respect, with mineral waters of medicinal value; such waters cannot be isolated from the rest of the ground water, but, because of certain special geologic-hydrogeologic conditions, are local accumulations of microcomponents and gases having medicinal value within that very ground water.

In addition, we should keep in mind that ground water also contains accumulations of oil in uneconomic quantities (or disseminated oil). In respect to this, Howell [1943] wrote that drops of free oil are frequently encountered in water in sandy rocks and that it is necessary to consider this circumstance in explaining both the origin of oil and the migration of oil. K. B. Ashirov [1949] has pointed out that all the rocks in the trans-Volga region are impregnated with disseminated oil drops, films of thick oil along fractures, rocks with the odor of petroleum, dark detritus-like inclusions of oil. G. M. Sukharev [1947] reported that samples of sandstones from beds XIII, XVI, and XXII on the southeastern anticline in the Oktyabr'skii region, outside the oil deposit, proved to be thoroughly impregnated with oil, but test holes yielded water with no indications of oil. A similar phenomenon has been observed in the Staro-Grozny region, at Tashkala, Oisungur, Syuil'-Kort, Makhach-Kala, Izberbash, and elsewhere.

Very interesting data on this matter have been obtained by K. B. Ashirov along Samarskaya Luka [1954]. In generalizing the data on the oil content of the rocks he noted that rocks showing strong impregnations of oil are frequently found on steep limbs below the oil-water interface, in places at great depths; these may represent traces of oil migration, forming a kind of train behind the deposit.

It is well known from exploration and exploitation practice that oil accumulates in so-called oil traps. But, natural traps for petroleum are hydrogeologic traps, representing a combination of geologic-hydrogeologic conditions that impede further migration of the oil and that favor the preservation of the oil deposit. Some of the hydrogeologic traps are due to marked changes in permeability (to oil and water) in the rocks forming the trap; others are found chiefly in domes, doubly plunging anticlines, and similar structures, related, as will be seen from what follows, to the sealing off of oil and gas that has separated from ground water.

From what has been said above it follows that oil fields are restricted to artesian basins or other types of pressurized-water systems, but that a single deposit is confined to a definite water-bearing horizon and, further, to that segment of the horizon where the geologic-hydrogeologic conditions are such that a hydrogeologic trap is formed.

In addition, it is necessary to discover if organic substances are present in ground water and, if so, in what quantities and of what kinds. *

It may be said, in general, that all ground water contains a certain quantity of organic material. But, whereas there is generally no doubt expressed concerning the presence of organic material in ground water, the amount and the composition of this material cannot be indicated with any certainty, since very little attention has been paid to organic substances in ground water; there has been, as yet, no proper and thorough study of this problem. Therefore, we may cite only some information taken from sources in the literature and some data obtained by us during investigations in the Grozny-Dagestan oil district. It is well known that ground waters in the north (the tundra and northern part of the forest-podzol zone) and in subtropical and tropical countries contain but very small quantities of humic material, and for this reason they even acquire a brownish or tannish color.

Table 7 shows some data on the content of organic material in mg per liter in ground water confined to rocks of various geologic ages. This information was borrowed from the well-known work of K. Keil'gak [1914].

TABLE 7

Content of Organic Material in mg per Liter of Ground Water

Shallow wells in rocks	Organic nitrogen	Organic carbon	Am- monia	Nitro- gen in nitrates and nitrites	Total nitro- gen	Total dry residue
Gneiss at Kendal	3.62	1.10	6.25	24.65	30.90	1002.0
Silurian at Alford	0.48	0.07	0	0.33	0.40	168.0
Devonian at Arbroath	1.68	0.64	0	41.97	42.61	1052.0
Devonian at Inverness	1.39	0.06	0	0.33	0.39	156.0
Carboniferous at Ogli-Hei	1.39	0.20	1.70	101.02	102.62	1207.0
Carboniferous at Durheim	1.24	0.45	0	62.68	63.13	1137.2
Carboniferous at Sheffield	12.00	1.26	1.10	0	2.17	185.0
Dolomite at Darlington	1.26	0.54	0.02	18.12	18.68	724.8
Sandstone at Newent	2.93	2.36	0	113.94	116.30	2321.2
Sandstone at Birmingham	3.40	1.05	6.20	147.17	153.33	2402.0
Sandstone at Newark	1.31	0.42	0	0.88	1.30	578.2
Liassic at Hillmorton	11.44	2.16	0.60	198.58	201.23	3068.5
Liassic at Semreton	8.04	1.43	0.05	94.49	95.96	1992.0
Liassic at Bitswell	2.05	0.33	0	0	0.33	1208.0
Oolitic at Thame	7.59	2.83	0.06	122.20	125.08	2696.0
Oolitic at Yorkton	4.48	0.56	0.05	1.78	2.38	490.0
Green sand and Walden clay at Pepper Harrow	0.14	0.12	0.01	67.22	67.35	714.0
The same, at Cambridge	0.79	0.27	0.74	0	0.88	859.6
Cretaceous limestone at Deal	2.41	0.34	17.00	63.45	77.79	1460.0
Cretaceous limestone at Marlborough	0.49	0.15	0	6.13	6.28	324.8
London clay	2.78	0.87	0	258.40	259.27	3965.0
London clay	2.87	0.76	26.50	54.34	76.89	1528.0
London clay	0.45	0.15	0.01	0	0.16	377.0
Alluvium at Whittlesea	9.31	9.40	30.50	108.48	143.00	2502.0
Alluvium at Windsor	0.15	0.10	0.01	0.63	0.74	300.8

The organic content of deeper, fresh, bicarbonate artesian water has always been studied merely from the sanitary-hygienic point of view. A large quantity of organic material has always been considered an indication of a close connection between any given ground water and the earth's surface and as an indicator of the degree of contamination. The amount of organic material is reflected in the quantity of oxygen, O_2, consumed, in mg per liter, corresponding approximately to 1 mg per liter of organic carbon. Table 8 furnishes some data on the content of organic and nitrogenous compounds in individual fresh-water aquifers of the Moscow basin.

* Organic substances associated with soil-forming processes and with the atmosphere are not considered here, since these are discussed in numerous papers in the literature devoted to studies of the soil.

The cited data refer to weakly mineralized waters having dry residues up to 1 g/liter and occurring at depths down to 200-300 m. V. M. Shvets, in treating the data on oxidizability of the Carboniferous waters in the Moscow basin, has shown that the Upper Carboniferous waters contain approximately 2.24 mg/liter organic carbon, the Middle Carboniferous approximately 1.98 mg/liter and the Lower Carboniferous about 2.22 mg/liter.

There is very little information on organic material in deeper sulfate and chloride waters.

TABLE 8

Content of Organic and Nitrogenous Compounds in Individual Fresh-Water Aquifers in the Moscow Basin

Aquifer	pH	N_2O_5, mg/liter	N_2O_3, mg/liter	Ammonia, mg/liter	Albuminoid ammonia, mg/liter	Organic material, mg of O_2
Upper Carboni-ferous	7.13-7.75	0.02-0.20	0.001-0.006	0.01-0.50	up to 0.2	1.6
Middle Carboni-ferous	7.7-7.8	0.01-5.0	0.01	0.2-0.7	up to 0.2	1.2
Lower Carboni-ferous	7.1-8.4	Commonly absent		0.02-1.17	Commonly absent	0.67
Devonian	—	Absent		—	Traces	0.8-9.48
Jurassic	6.5-7.4	Absent		—	Traces	0.21-3.0
Cretaceous	4.8-6.8	Absent		Traces	—	0.8-1.5

We should note first that, according to the studies of Hackford [1932], water (associated with oil deposits) taken from wells gives a reaction to sugar and forms osazone with phenylhydrazine acetate.

According to V. G. Malyshek [1040], the waters associated with oil deposits on the Apsheron Peninsula contain one gram of organic material per 100 g of water, or 10 kg per m^3 of water. This material consists of sodium salts of organic acids, naphthenic acid being most abundant; but fatty acids are also present. The amount of naphthenic acids decreases with increasing distance from the boundary of the oil deposit. B. G. Malyshek explains the appearance of sodium salts of organic acids by the reaction between alkaline waters and petroleum, as shown in the equation

$$RCOOH + NaHCO_3 = RCOONa + H_2O + CO_2.$$

According to F. F. German [1951], waters at deposits in the red-beds of Western Turkmenia contain the quantities of organic substances shown in Table 9.

The organic material in the upper horizons consists almost entirely of unsaturated (naphthenic) acids. With increasing depth, the quantity of saturated (fatty) acids increases, but naphthenic acids are still clearly the dominant form. In regard to this, F. F. German expressed his opinion that petroleum is enriched in paraffins by formation paraffins from naphthenic and fatty acids (introduced by water) through the effect of alpha radiation. He adds, further, that fatty (saturated) acids, although not susceptible to reduction, may decompose under the influence of alpha radiation according to the following scheme:

$$RCOOH \rightarrow RH + CO_2,$$

and yield paraffin hydrocarbons.

To this it may be added that the presence of naphthenic acids has been ascertained in a very great number of waters associated with oil deposits.

According to V. A. Sulin [1946], the content of naphthenic acids in waters associated with oil deposits in the USSR ranges between wide limits: from negligible quantities to hundreds of milligram equivalents per liter. These data show that calcium chloride and magnesium chloride waters contain smaller quantities of naphthenic acids then alkaline waters, a fact related to the poor solubility of calcium salts in naphthenic acids. For example, the calcium chloride waters of Vtoroi Baku contain less than 0.1 milligram equivalent organic acids per liter; the magnesium chloride waters of Buguruslan contain more, up to 0.2-0.4 milligram equivalent per liter. On the other hand, bicarbonate waters in the Caucasian oil province contain naphthenic acids in amounts up to several tens of milligram equivalent per liter, and the oil itself contains from 0.1 to 2.4% naphthenic acids.

TABLE 9

Content of Organic Material in Waters at Deposits in the Red-Beds of Western Turkmenia

Designation of water	Depth to water from roof of red-beds, m	Specific gravity	Content of organic material, mg/liter
Water from roof and upper horizons of red-beds	0-150	1.174-1.86	2.2-2.8
Ferruginous water	100-200	1.160-1.64	None
Water from upper horizons	290	1.138	3.6
Water from middle horizons	512	1.100	4.2
Ditto	774	1.020	10.6
Water from lower horizons	1040	1.018	20.2
Ditto	1127	1.016	28.9

Our own investigations in the Grozny district, in ground water in the Chokrak and Karagan strata, revealed 3.4 mg/liter of organic carbon in the regions of recharge (Chernye Mountains); in the region of oil deposits (Peredovye Ranges), waters in wells having a yield of over 300 m^3 per day proved to have 6.6 mg/liter of organic carbon, whereas waters in wells having a yield less than 300 m^3 per day were found to contain 34.7 mg/liter of organic carbon; in regions of discharge (thermal springs of the Peredovye Ranges) the content of organic carbon was found to be 3.4 mg/liter. More detailed data on the Grozny district are presented in Table 10.

In the Dagestan district, at sites of recharge for the Chokrak and Karagan strata, the content of organic carbon is somewhat less, 1.82 mg/liter (average of 16 samples); in areas of oil deposits and in exploratory fields (where wells yield only water, the content is 1.53 mg/liter (average of 14 samples); in areas of oil deposits where wells yield both oil and water the amount is 19.57 mg/liter (average of 16 samples); and in zones of discharge, the content is 5.11 mg/liter (average of 16 samples). Waters of the Khadum and Lower Cretaceous deposits in the region of gas deposits contain 5.93 mg/liter organic carbon (average of 11 samples).

These data lead to the following conclusions:

a) organic carbon is everywhere present in ground water;

b) the quantity of organic carbon increases away from regions of recharge and is especially large at zones where the movement of the ground water is curtailed (low-yield wells at oil and gas deposits);

c) in the Dagestan district, the organic carbon is less where the oil content is less. The organic content of carbon is even less where the oil content is less. The organic carbon content is even smaller in nonpetroleum-bearing waters in the Carboniferous rocks of the Moscow basin.

Our investigations in the Grozny-Dagestan petroleum district have established the fact that naphthenic acids are widespread in ground water. Some of the summary data are as follows: the content of naphthenic acids is 0.11 mg/liter in recharge areas (average of 27 samples); in areas of oil deposits where wells given only water, the content is 0.24 mg/liter; in wells giving both oil and water, the content is 1.55 mg/liter; in areas of gas deposits, the content is 0.27 mg/liter; and at places where ground water is discharging, the content is 0.95 mg/liter. These data indicate that naphthenic acids do not occur only in the marginal zones of oil or gas deposits, as had previously been thought. Actually, they are present even at sites of ground-water recharge (in the Chernye Mountains, for instance), where they could not possibly occur if they were decomposition products in oil deposits. In nature they obviously precede the formation of hydrocarbons.

A. F. Dobryanskii, for example, has stated that naphthenic acids are associated not with the principal mass of petorleum, but with some component of the source material. Various sterols and vitamins, especially Wieland's nucleus of "D" (for cholesterol), are widespread in nature. No less interest has been aroused by camphene and

TABLE 10

Content of Organic Carbon in Ground Water from Productive Beds in the Grozny Oil District[1]

Region of recharge (Chernye Mountains)			Region of the deposits (Peredovye Ranges)						Region of discharge (Peredovye Ranges)		
			Group I (water from oil-free wells or from wells gushing oil or yielding more than 300 m³ per day)			Group II (water from wells with oil, but yielding less than 300 m³ per day)					
age	sample no.	C_{org}, mg/liter	age	sample no.	C_{org}, mg/liter	age	sample no.	C_{org}, mg/liter	age	sample no.	C_{org}, mg/liter
Chokrak	12	2.8	Chokrak	Oil-free wells		Chokrak	5	81.0	Chokrak	22	2.1
	13	1.5		44	6.5		17	27.8		20	2.4
	69	—		26	1.4		1	55.0		21	2.0
	83	4.2		35	3.8		9	40.6		32	2.6
	84	2.3					43	12.1		27	3.2
			Karagan	41	3.1		45	13.0		28	2.0
				74	16.0		42	37.4		29	2.9
Karagan	72	4.4					46	25.4		30	2.2
	70	5.0					48	13.5		49	3.0
							47	16.2		67	2.9
							37	45.2		68	2.9
							24	20.1		63	5.6
Ave. 3.4 mg/liter or ~ 7 g/m³			Chokrak	Gushing and high-yield wells			51	5.3		59	5.2
							50	11.5		62	2.3
				4	9.8		75	12.0		55	6.7
				18	3.4		76	19.5			
				23	5.8		77	40.0			
				25	4.2		78	27.3	Karagan	64	6.4
							80	35.0		58	3.5
							79	32.8		57	5.2
				10	10.2					56	3.7
				2	5.6					82	—
			Karagan	6	9.7	Karagan	16	38.7		33	4.4
							15	43.2		34	1.3
							7	27.5			
							8	163.0			
							14	11.4	Ave. 3.4 mg/liter or ~ 7 g/m³		
			Ave. 6.6 mg/liter or ~ 13 g/m³				3	23.7			
							38	33.3			
							36	70.6			
						Ave. 34.7 mg/liter or ~ 70 g/m³					

[1]Data for 1954.

isocamphene, the presence of which might supply the typical nucleus of naphthenic acids. In Dobryanskii's opinion, the participation of bicyclic terpenes might connect naphthenic acids with the plant kingdom.

From all this it may be concluded that naphthenic acids have a direct genetic relationship with petroleum. It is true that some investigators (V. G. Malyshek, V. A. Sulin, and others) believe that the abundance of naphthenic acids in waters associated with oil deposits is due to the derivation of these acids from the oil, but, contrary opinions are to be found, maintaining the possibility that the hydrocarbons of oil are derived from naphthenic acids. Thus, T. L. Ginzburg-Karagicheva has proposed such a scheme of alteration by biogenic processes of the initial organic substances in oil:

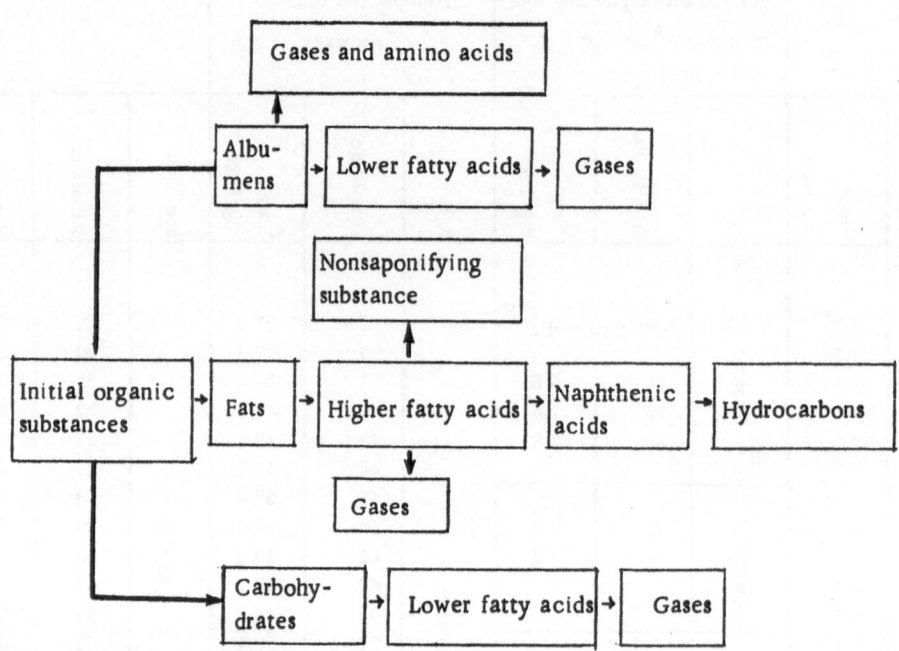

From the scheme of T. L. Ginzburg-Karagicheva it follows that albumens and carbohydrates are ultimately converted to gases, but fats yield naphthenic acids, which, in her opinion, may presumably be converted to hydrocarbons.

Out data obviously supports precisely this point of view relative to naphthenic acids.

To all this it is necessary to add that it has long been known that hot springs in the Grozny region, in the areas where the Produktivnaya (Productive) series discharges, bear a film of liquid petroleum. This fact is generally explained as being due to gradual destruction of oil deposits. To support this explanation it should be demonstrated that the oil deposits of the Grozny region are actually being destroyed by the mechanical activity of ground water. However, there is no basis for such a belief. Further, it would be necessary to prove that there is a flow of ground water from an oil deposit being destroyed to the point of emergence of a hot spring. But, this possibility is contrary to the view that there is an extensive zone of hydraulically stagnant water near anticlinal crests with hydraulic closure, or near shielded deposits; it is also opposed to the concept that ground water does not flow away from a deposit, but, on the contrary, supports the deposit, creating a certain formational pressure in it and in the gas cap.

In addition, as experience at the Staro-Grozny oil fields has shown, the Goryachevodskaya springs, which are closely connected hydraulically with the oil fiels, have now gone dry as a result of pumping at the wells. But, these springs did now show oil films as they once did. On the other hand, the springs at Bragunskiye and Isti-su, which are some distance from pumping wells and are not near any oil deposit, still bear oil films, although the the direction of flow of the ground water has changed fundamentally because of the pumping. The water now flows toward the oil field, and the oil films at the Bragunskiye springs cannot possibly be due to destruction of oil deposits in the Staro-Grozny region.

38

Z. A. Tabasaranskii [1954] has reported that luminescent-bitumen studies of ground water in the Il'sko-Kholmskii region have revealed particles of liquid hydrocarbons. In connection with these studies, Z. A. Tabasaranskii is inclined to believe that the process of oil formation is at present in operation, but, in a footnote, he suggests that the phenomenon may also be explained by the destruction of oil deposits. But again, there is no proof offered for this latter explanation.

In general, this is still rather meager information concerning the organic material in ground water; but, firstly, it points out, above all, the necessity of thorough studies on such material in the various hydrochemical zones, and, secondly, it does not deny but, on the contrary, it confirms the possibility of the accumulation of such material and its conversion to oil as a result of a series of physicochemical and biochemical processes.

It is possible to say with complete assurance that any sort of broad, regional study of the organic material and microflora in potential and known oil regions will undoubtedly produce important data, of great value not only in seeking a solution to the problem of the origin of oil but also for treating intelligently the hydrochemical criteria for evaluating oil districts and for prospecting for oil and gas.

From the point of view of balance, the organic substances in ground water are more than sufficient to produce both commercial deposits and disseminated oil. More likely, not all the organic material present in ground water is used for forming oil, but only part of it, since the formation of oil, as will be pointed out below, requires a series of specific conditions not everywhere present.

According to V. A. Sokolov [1948], if the industrial reserves of petroleum are taken to be 2×10^{10} tons, the average yearly accumulation of oil throughout the earth amounts to some tens of tons, whereas, if this quantity is referred to areas where sedimentary rocks are distributed, we obtain a figure amounting to only several grams per km^2. This quantity is insignificant in comparison with the quantity of organic material that takes part in soil-forming processes alone.

Oil deposits represent a very small part of the total volume of water-bearing strata containing them. If we take a zone of water-bearing formation to have a maximum width of only one kilometer, if it extends for 100 km, is 50 m thick, and has a porosity of 20%, the volume of water contained in the zone will be 1,000,000,000 m^3, and this will contain approximately 10,000,000 tons of organic material, if we assume the organic content to be 10 kg per m^3 of water (according to V. G. Malyshek). Furthermore, it is necessary to keep in mind that ground water, however slowly, continues to move with some velocity; i.e., in the course of time it is renewed. Thus, slowly moving ground water, in the course of geologic time, draws after it a tremendous, continuous stream of organic material, which goes into the formation of oil.

The water-exchange factor in depressed geologic structures indicates that the volume of ground water occurring in these structures is completely replaced after a certain interval of time. From this it follows that the quantity of organic material contained in the ground water before the formation of oil deposits may be computed by multiplying the instantaneous content of organic material by the time interval between the formation of the water-bearing horizon (i.e., practically the time represented by its geologic age) and the time the oil deposit developed, and then dividing this value by the water-exchange factor.

The above-cited facts support the view that ground water everywhere contains organic material in greater or lesser quantities. The quantity contained is entirely sufficient for the formation of the known reserves of oil. Nor is there any doubt that the organic substances, migrating through water-bearing strata, undergo various transformations because of salts and gases in the ground water or because of the life activities of microorganisms. But, perhaps these changes have no direct connection with the processes of oil and gas formation. In the light of our existing knowledge about the Grozny-Dagestan oil region, it is not possible to solve this problem to our complete satisfaction.

The chief value of the data lies in its proof of the necessity of studying organic substances and microflora in ground water in connection with the processes of oil formation. Therefore, we shall examine below the processes already discussed in the literature, processes leading to the formation of this or that component of natural petroleum, and we shall evaluate the possible effectiveness of these processes in the subsurface hydrosphere.

Let us pause first on physicochemical processes.

According to A. F. Dobryanskii [1948], methane hydrocarbons may be derived from the decomposition of wax-like compounds and, in general, esters, anhydrides of acids, and acids themselves or other nonhydrocarbon compounds containing the long methane radical. The aromatic hydrocarbons should be considered substances that form synchronously with oil and that are derived directly from the initial source material. The course of alteration to oil from aromatic to naphthene and, lastly, to methane is the result of isolating the single-ring aromatic hydrocarbons from the more complex, polycyclic forms. Naphthenic acids are not found in large quantities with paraffins; their low content indicates that they are genetically related not with the principal mass of petroleum but with some component of the source material. Sulfur compounds are due to introduced sulfur, hydrogen sulfide forming by biochemical reduction of sulfates in the presence of organic material; biochemically, hydrogen sulfide breaks up on the formation of sulfides, and only part of the sulfur is taken up by the oil, in the form of complex and unstable compounds. Nitrogenous substances form from albumen; the tarry content of petroleum is associated with profound regeneration of the oil, involving the development of cyclic compounds, or else, in the process of alteration in the oil, tarry substances may disappear as the number of cyclic compounds decreases.

The source of hydrogen imbalance, according to A. F. Dobryanskii, is internal heat, and the cause of the heat is found in catalytic processes.

The chemical aspect of A. F. Dobryanskii's hypothesis on the origin of oil, the principal idea of which reduces to the redistribution (imbalance) of hydrogen, can hardly be accepted, since the experiments of V. I. Ipat'ev have shown that hydrogenation occurs first, and then, at very high temperatures (on the order of 400°C) the initial substances begin to break up and hydrogen is redistributed. The view of A. F. Dobryanskii concerning the formation of methane hydrocarbons from wax-like compounds, anhydrides of acids, and acids themselves is a denial of the hypothesis of oxidation of naphthene hydrocarbons to naphthenic acids; the formation of sulfur and nitrogen compounds as suggested by Dobryanskii deserves our full attention, since a number of the processes suggested by him actually take place in ground water, or, at least, their experience is very probable.

According to G. L. Stadnikov [1937], oil is formed by the process indicated below.

1. Hydrogenation of decomposition products of plant material at comparatively moderate temperatures may produce a mixture of substances very similar to oil in composition and properties.

2. Unsaturated fatty acids are converted, by polymerization, to high-molecular cyclic compounds, which are easily dissolved in the acid that has not begun to polymerize. Waxes and tars should also dissolve in this new solution, and humic material with them. Oxidation is impossible in a brackish-water basin with hydrogen sulfide fermentation, but polymerization of unsaturated acids occurs very slowly. Some of the polymerized compounds are converted, in the presence of anaerobic microbes, to hydrocarbons or to ketones, and they lose CO_2; part of the nonpolymerizing saturated acids also undergoes such alteration. The polymerized compounds and the hydrocarbons, together with the tars and waxes, being dissolved in the mass of fatty acids, develop an environment in which humic substances have been dispersed. Primary petroleum has formed in this manner. This primary petroleum consisted of a thick, tar-like mass, having high viscosity and a specific gravity greater than one (otherwise the petroleum would be able to float on water).

3. The substances continue to change slowly, the processes involving partly the polymerization of unsaturated acids and the formation of cyclic polymers and partly the decarboxylization of simple and polymeric acids.

4. Then gases (H_2 and CO) reach the "primary petroleum" from the interior of the earth; these gases result from reactions between water and incandescent solutions of metallic carbides, and they hydrogenate unsaturated and oxide compounds. This explains the formation of mixtures of hydrocarbons in the methane and naphthene series.

5. If large quantities of saturated acids in the fatty series accumulate in the material, but little humic material or tar, then methane petroleum poor in tars is obtained during hydrogenation.

In this connection, methane oils are accompanied by (in addition to hydrocarbon gases) a still greater quantity of hydrogen (Pittsburgh, Pennsylvania — from 7 to 35%), whereas CO_2 is almost absent. On the other hand, aromatic oils yield considerable CO_2 and have almost no hydrogen.

When unsaturated fatty acids are dominant and there is little admixture of humic material, a mixture of obtained that consists of chiefly polymeric high-molecular acids; on hydrogenation, a hydrocarbon mixture is produced that is dominated by naphthenes. When unsaturated fatty acids are present and the accumulation of humic material and tars is large, the fatty acids and humic materials are polymerized to form cyclic polybasic acids, but the tars, on contact with air, are oxidized and polymerized, a process leading to the conversion of some of them to high-molecular, relatively insoluble substances similar to asphaltenes. The result is a primary petroleum containing few saturated fatty acids, large quantities of cyclic acids, much humic material with aromatic structure, tars, and asphaltenes. Hydrogenation of such a mixture leads to the formation of petroleum of the naphthene-aromatic class.

Thus, according to G. L. Stadnikov, the fundamental process of oil formation is hydrogenation, which facilitates processes of polymerization and decarboxylization of simple and polymeric acids. In his view, the quantitative relations of the source material should affect the quality of the future petroleum.

The hypothesis of G. L. Stadnikov is considerably nearer the truth of what actually occurs in ground water. The preference shown by him for the more stable remains of the plant kingdom (waxes, hydrocarbons, tars, sporanins, pollenins, stearins, cutins, fatty acids, and lignins) is very sound, since it is precisely these constituents that charge the ground water after destruction of the unstable components of the plant material; they form in an aerobic environment on the earth's surface and, in part, in the soil cover. But, in this regard, there is something of a contradictory note in one of the basic tenets of this hypothesis, and that is that all the source material of all petroleum is thought to consist of fats of vegetable origin and not of all the fundamental complex of stable residues. It seems to us, as to all geologists, that this represents an almost unbelievable accumulation of large homogeneous masses of fatty material, whereas this is actually contrary to the broad, regional distribution of oil deposits and of disseminated oil within the earth's crust.

The views of G. L. Stadnikov are very important in regard to hydrogenation; to the conversion of unsaturated acids to cyclic compounds; to the solution of tars, waxes, and humic material in these compounds; and to the fact that saturated fatty acids lead, above all, to the formation of methane oils, the presence of large quantities of tar and humic material to aromatic oils, and the presence of unsaturated acids in naphthene oils.

A view of G. L. Stadnikov that appears extremely unlikely is the one proposing the introduction of hydrogen from the deep interior of the earth's crust by reaction between water and incandescent metallic carbides (already discussed above under the evaluation of the hypothesis of an inorganic origin of petroleum). The idea of primary petroleum is also improbable — a thick tar-like mass — since such a concept leads one to deny the migration of oil (no one having ever observed the movement of such thick, tar-like material) and to contradict basic opinions on the hydrodynamics of liquids and semiliquids in the earth's crust.

A. V. Frost, in a joint paper with V. A. Geitling [1945], in alluding to experiments of Zelinskii showing that Al_2Cl_3 converts fatty and tarry substances to oil-like products at high temperatures, wrote that if we assume the initial material to be cellulose, two variants are possible: first, when the bacteria in an anaerobic environment of a closed marine basin convert cellulose to alcohols, ketones, and fatty acids and, acting on the fats, convert them to acids (partly decarboxylizing), which afterward, in the presence of clay catalysts and at higher temperatures, are converted to petroleum; and secondly, when cellulose and fats, at high temperatures and in the presence of alkaline rocks, marine water, and clay, are converted to a mixture of alcohols, ketones, and salts of fatty acids, which then are converted to oil or gas by the effect of catalyzing aluminosilicates. The experimental data of A. V. Frost have shown that, at 300°C, cellulose, regardless of the length of time it is treated, yields no primitive (proto) oil, but in the presence of clays and alkalis these reactions take place at temperatures as low as 250°C.

In one of his latest papers, written jointly with L. K. Osnitskaya [1951], A. V. Frost has reported that, at temperatures below 260°C, a number of transformations occur: a) cellulose, acted on by bacteria, yields acids, alcohols, and ketones, which change to paraffin hydrocarbons; b) fats and waxes, acted on by bacteria, are saponified and, in the presence of clay, form paraffin and ceresin; c) tars, stearins, and terpenes, in the presence of catalytic clays, yield naphthenic and aromatic hydrocarbons; d) proteins become the base of nitrogenous compounds; and e) lignins are transformed but the manner is not clear.

Work in the petroleum laboratory of the All-Union Geological Scientific-Research Institute (A. I. Bogomolov and K. I. Panina) has shown that in the presence of clays and at moderately high temperatures methane, naphthene, and aromatic hydrocarbons of various molecular weights may be assumed to form from fatty acids.

Very interesting experiments have been made by V. N. Ipat'ev [1936] in studying the common course of the processes of catalysis and hydrogenation. In this case, the experiments were the more important because the catalytic reactions were studied not only with dry powders, but also in aqueous solutions. We shall therefore dwell in some detail on the results of the experiments and on the conclusions of V. N. Ipat'ev.

In summarizing his experiments on the hydrogenation of salts of aromatic acids in aqueous solutions, Ipat'ev has written that the formation of hydrocarbons during hydrogenation occurs both with dry salts of the acids and in solutions, only the generation of hydrocarbons is considerably greater in the latter case. In addition, when hydrogenating soluble salts, the acids obtained are completely hydrogenated. The reaction proceeds according to the equation

$$2RCOONa + 4H_2 = 2RH + Na_2CO_3 + H_2O$$

$$C_6H_5COONa + 3H_2 = C_6H_{11}COONa$$

Water fosters both reactions.

Several of V. N. Ipat'ev's experiments are outlined below.

Experiment with sodium acetate.

A solution of sodium acetate, being heated with aluminum oxide to 300°C, gives up to 40% methane in the presence of H_2.

Experiment with sodium benzoate.

50 g of this salt in 75 cm^3 of water with 7 g of nickel oxide and in the presence of H_2, at a pressure of 90 atm and a temperature of 300°C, gives 9.7% CH_4, 90% H_2, and 9 g of cyclohexane, having a boiling point of 79-80°C.

Experiment with sodium amygdalate.

50 g of this salt in 75 cm^3 of water with 7 g of nickel oxide and in the presence of H_2, at a pressure of 80 atm and a temperature of 275-286°C, yields a mixture of hydrocarbons, the bulk of which consists of hexahydrotoluene with a small admixture of toluene.

Experiment with sodium enanthate.

25 g of this salt in 70 cm^3 of water with 3 g of nickel oxide and in the presence of H_2, at a pressure of 70 atm and a temperature of 315°C, gives methane and hexane with a boiling point of 68-72° (according to the equation below):

$$2CH_3 - (CH_2)_5 - COONa + 4H_2 \rightarrow 2C_6H_{14} + CH_4 + Na_2CO_3 + H_2O.$$

According to the experimental data of V. N. Ipat'ev, the generation of hydrocarbons with boiling points of 106-110°C and 130-135°C occurs with sodium phenylacetate and sodium cinnamate.

In a number of his experiments on hydrogenation he succeeded in going directly from terpenes to hydrocarbons. He concluded that the hydrogenation of salts of aromatic acids in aqueous solutions proceeds in a different manner from hydrogenation of dry salts; in addition to hydrogenizing, the acids decompose, with the separation of hydrogenized hydrocarbons. He has noted that, in this process, double-bond compounds hydrogenize better than the aromatic ring.

The following results may be noted from other experiments of V. N. Ipat'ev.

1. Various sodium salts of organic acids give hydrocarbons with different boiling points. For example, sodium enanthate gives a hydrocarbon with a boiling point of 68-72°C, sodium benzoate, a hydrocarbon with a boiling point of 79-80°C, sodium phenylacetate 106-110°C, and sodium cinnamate 130-135°C.

2. When aromatic acids are being hydrogenized in aqueous solutions, an additional amount of methane forms.

The catalyst nickel, as is well known, constitutes one of the chief elements in petroleum ash.

A by-product of hydrogenation in aqueous solutions is Na_2CO_3, which is very clearly related to the alkalinity of many waters associated with petroleum deposits.

In examining the experiments of V. N. Ipat'ev, perhaps it should be said that the processes of hydrogenation and catalysis reproduced by him occur at high temperatures (on the order of 300-400°C), and this is in contradiction to the almost solidly established position that oil is formed at considerably lower temperatures (on the order of 75-100°C). But, lowering the temperature of reaction, in most cases, leads merely to slowly down the process in accordance with the well-known law of physical chemistry that a decrease in temperature of 10° reduces the rate of chemical reaction by one-half to two-thirds. A decrease of 200° in the temperature, of course, may possibly slow the reaction by a factor of several thousand. However, in nature, oil does not form instantaneously, but the process is drawn out over hundreds of thousands and even millions of years. In addition, we must consider that the experiments of A. F. Frost and his coworkers prove empirically that, for hydrogenation and catalysis of hydrocarbons in general, such high temperatures are not required. The temperature, in his opinion, may be lowered to 150°C and more; and the pressure may be as little as 10-20 atm (according to V. N. Ipat'ev).

V. N. Ipat'ev used nickel oxide as a catalyst in his experiments. But, the same processes may be effective with other inorganic catalysts. For example, as early as 1901, Sabatier and Senderens established the fact that terpenes yield either naphthene or aromatic hydrocarbons in the presence of aluminum oxide. At the present time, it is already known that ground water is very rich in various microcomponents, including well-known catalysts such as aluminum, nickel, copper, zinc, and many other elements.

The following other interesting conclusions of V. N. Ipat'ev should also be noted.

1. Alcohols, at 360°C, ordinary pressures, and in the presence of a clay catalyst, decompose to olefins (ethylenes) and water. An increase in pressure merely slows the rate of the reaction.

2. Aromatic aldehydes, in the presence of a nickel catalyst, are hydrogenated to aromatic hydrocarbons and, in part, to toluene.

3. Metal-organic compounds in a dry form, during hydrogenation, decompose with the separation of the metal and with the formation of a corresponding hydrocarbon.

4. Silicon-organic compounds are also hydrogenated.

V. N. Ipat'ev performed very interesting experiments with naphthenic acids. The heating of these acids leads to cracking in three directions: a) with the splitting off of the carboxyl group and with the formation of hydrocarbons, b) with the splitting off of side hydrocarbon chains, leading to the formation of naphthenic acids of lower molecular weight, and c) with the splitting off of side chains containing the carboxyl group, leading to the formation of fatty acids.

Hydrolysis of caustic soda solutions of naphthenic acids leads to the formation of benzenes, kerosenes, and a solution of soda with admixtures of fatty acids.

At atmospheric pressure, naphthenic acids yield olefins, but at high pressures they give hydrocarbons of the paraffin and naphthene series, as shown by the following experiment: 100 cm^3 of naphthenic acid in 50 cm^3 of water, after five hours in the presence of 10 g Al_2O_3 at a temperature of 440-460°C, yields in the water-free residue a mixture of hydrocarbons of the paraffin and naphthene series, and an admixture of aromatic compounds. According to V. N. Ipat'ev, water inevitably participates in the catalytic reaction, oxidizing metals during the decomposition. S. S. Nametkin has shown that naphthenic hydrocarbons are derived from essential oils and aromatic hydrocarbons by catalytic hydrogenation. Naphthenic hydrocarbons are hydrogenated to saturated hydrocarbons.

In general, it may be stated, in complete agreement with a great number of investigators concerned with the problem of the origin of oil, that hydrogenation and catalysis, especially in water-bearing horizons (i.e., in an aqueous environment), may be fundamental factors in the formation of the liquid components of oil. The investigations of V. N. Ipat'ev show that the initial source material in water-bearing horizons, during these processes, may be alkali salts of organic acids, and, above all, aromatic and naphthenic acids.

In the light of these experiments a number of considerations of N. T. Shabarova are also of interest; Shabarova assesses considerable importance to the salts of organic acids as source material in the process of oil formation. For example, she has suggested [1954] that hydrolysis of lipides, albumins, and carbohydrates leads to the formation of various intermediate products: aldehydes, alcohols, organic acids, amines, ammonia, and other compounds. Some of the compounds break down further to gaseous substances (carbon dioxide, water vapor, hydrogen sulfide, molecular nitrogen) and are returned to the atmosphere. In addition, she has noted that organic acids easily react with inorganic elements, forming salts and complex compounds. In an alkaline solution, at high pressures and temperatures and in the presence of haloid derivatives and sulfur compounds, the salts of organic acids may be reduced to hydrocarbons. The role of nitrogenous substances in processes of oil formation may be summarized in a phrase: they protect organic compounds in the sediments from oxidation.

Let us now consider radioactive processes.

V. A. Sokolov, in a paper devoted to the origin of oil [1948], considered in detail the significance of radioactive processes. He noted that radioactive substances in sedimentary rocks are most abundant in clays, less abundant in sandstones, and even less abundant in limestones. According to Duane, one curie of radium emanation forms about 70 cm^3 of oxyhydrogen gas ($2H_2 + O_2$) in water per day. According to V. A. Sokolov's computations, radioactive waters associated with oil deposits, after one million years, form from 1 cm^3 of water 0.0075 cm^3 of hydrogen; and 10,000 tons of H_2 may give rise to 250,000 tons of petroleum in 1 km^3 of rock; Tertiary deposits with an age of 10 million years might yield 2.5-3 million tons of petroleum per km^3 of rock. However, according to observations of V. A. Sokolov himself, radioactive processes, if the oil forms in clay rocks, can provide for the formation of only 1% of the existing oil reserves (such as in the Pliocene clays of the Baku region). However, if oil migrates throughout an extensive region in thick reservoir beds and if only afterward it is confined to the crests of anticlines, radioactive processes, in the opinion of V. A. Sokolov, may be of fundamental importance in the formation of oil.

The investigations of V. A. Sokolov make apparent that, of three types of distribution of radioactive elements (uniformly distributed in water, spread on the surfaces of particles constituting rocks, and contained in particles of the rock), the radioactive elements distributed in water produce the greatest effect of alpha radiation (Fig. 4). Actually, the length of path of an alpha particle from RaC' is 0.06 mm. Consequently, for particles having diameters greater than 0.12 mm, the total effect of alpha radiation will be absorbed within the rock, and decomposition of radioactive elements on the surface of particles will supply but 50% of the alpha radiation to water; 50% will be directed into the mineral particles.

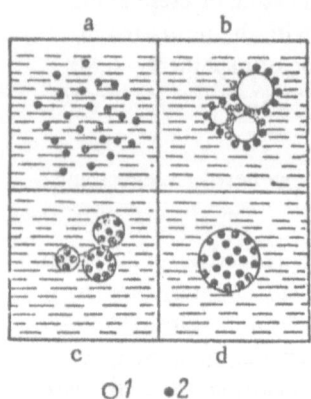

Fig. 4. Various types of distribution of atoms of radioactive elements (after V. A. Sokolov). a) Uniform distribution of atoms of radioactive elements in water, b) the same number of atoms, but distributed on the surfaces of particles in the rock, c and d) the same number of atoms, but distributed uniformly inside the particles of the rock, 1) particles of the rock, 2) sphere of activity of the alpha particles.

In conclusion, V. A. Sokolov has acknowledged that, although alpha radiation undoubtedly has some effect on the formation of oil, it cannot be a decisive factor. In his opinion the most important factor is the slow thermal decay of hydrogenous and humic substances at moderate temperatures. He also has recognized the necessity, in the formation of oil, of hydrogenation by hydrogen that formed through thermal decay and dissociation of water by alpha radiation.

An increase in the generation of gases, including combustible gases, during fermentation of organic material in the presence of radon, is indicated by Zheslen's data, shown in Table 11.

According to V. A. Sokolov, when fatty acids are exposed to alpha radiation, H_2, CO_2, CO, CH_4, and liquid hydrocarbons are formed.* Other investigators have shown that a mixture of saturated and unsaturated hydrocarbons is formed when naphthenic acids are exposed to the radiation of alpha particles.

The wide, regional distribution of radioactive elements (10^{-10} to 10^{-13}%) in ground water (the concentration being high locally even in underground flow near the earth's surface), the increase in quantity of such elements with depth and the rather high concentration in deep, highly mineralized waters (including waters associated with oil deposits, 10^{-10} to 10^{-12}%) impel us to acknowledge that the decomposition of molecules of ground water through alpha radiation actually represents a source of hydrogen generation, the element obviously necessary for hydrogenation, and to recognize the fact that this process undoubtedly plays some part (in places of high concentrations even a very essential part) in the formation of petroleum.

*Collection: The Chemical Effect of High-Energy Radiation, Foreign Literature Press, 1949.

TABLE 11

Gases Obtained During the Fermentation of Organic Material (after Zheslen)

Gases	Gases, obtained in a year's time, in milligrams, for equal quantities of organic material	
	in the presence of radon	in the absence of radon
N_3	372	142.7
CO_2	320.4	242.1
Combustible gases	409	410.8
Totals	1101.4	795.6

In regard to the temperatures and pressures necessary for the formation of oil, there is now almost unanimity of opinion, this being that the temperature must be moderate, the pressure high. Some brief remarks of individual authors on this matter are quoted below.

V. P. Baturin [1945] and many others maintain that we may now consider as a proved fact the idea that temperatures in the zone of oil formation could scarcely have exceeded 100-150°C, and the presence of living microflora suggests, apparently, that the temperature must have been even lower than 100°C.

P. I. Walden believes that pressures should be high, but not temperatures, during the formation of oil.

A. F. Dobryanskii has pointed out that the temperature should be low (125-130°C) but that the pressure should facilitate the reaction, changing with decrease in volume (for example, hydrogenation by introduced hydrogen).

A. V. Frost believes that the Baku petroleum was formed at temperatures no higher than 90°C.

According to V. A. Sokolov, geologic conditions indicate that oil is formed at temperatures not exceeding 100°C, even at 50-60°C and lower.

Let us move now to microbiological processes.

Most investigators grant some significance to the life activities of microflora in the formation of petroleum. Some even believe that the main bulk of hydrocarbons developed in the cells of various bacteria, such as desulfurizing bacteria. But, the opposite point of view is also held. For example, A. S. Uklonskii [1940] believes that discoveries of microflora are due to the impossibility of obtaining sterile samples.

In V. A. Sokolov's opinion it is most likely that bacteria have nothing to do with the formation of such typical hydrocarbons as C_2-C_8. G. L. Stadnikov also believes that the role of bacteria is uncertain. However, he has noted that opinions about oil, as about antiseptics containing phenol, cannot be considered definitely proved.

T. L. Ginzburg-Karagicheva [1940] believes that the extensive occurrence of microspores leads to the reduction of sulfates. She has noted, further, that data on bacterial conversion of fatty substances, on the one hand, and of naphthenic substances, on the other, allow us now to observe a bridge — naphthenic acids — connecting the initial fatty substances with the final products, petroleum hydrocarbons.

In one of her most recent papers, T. L. Ginzburg-Karagicheva [1953] has reported a summary of data from her investigations, outlined below:

1) microbes which reduce sulfates, with the formation of H_2S and FeS, have been discovered as constituents of microflora;

2) there are microbes that actively decompose various organic substances, with the formation of gaseous products (CH_4, H_2, CO_2, NH_3, and H_2S) and, as indicated by a number of investigators, higher hydrocarbons;

3) microbes have been separated and studied which produce profound changes in fatty substances, with phenomena of polymerization, hydrogenation, and decarboxylization of higher fatty acids, and with the formation, in several experiments, of dark products having a marked bituminous odor;

4) microbes have been identified in petroleum which change naphthenic acids to unsaturated naphthenic hydrocarbons in an anaerobic enrivonment.

These general conclusions, it seems to us, have a fundamental significance for solving the problems concerned with the origin of oil.

Considering the fact that sterile ground water, at least in all the upper hydrogeologic zones (except for steam-hydrothermal), does not exist and that ground water in all geologic epochs has been closely connected to the earth's surface through zones of recharge, the surface being inhabited by large numbers of various microflora, we have included the following zones in our microbiological investigations in the Grozny region: the zone of recharge (exposures of aquifers in the Chernye Mountains), the zone where ground water has been sealed off by oil deposits (waters associated with oil deposits in the Terek and Sunzha ranges), and, lastly, the zone of discharge of ground water (hot springs in the Grozny region).

All these indicated zones contain various microflora: denitrifying, desulfurizing, methane-forming (in environments of carbon dioxide and hydrogen and of ethyl alcohol and calcium acetate), cellulose-destroying, methane-oxidizing, heptane-oxidizing, oxygenizing, and thionic acid-producing (data of Z. I. Kuznetsova).

In the recharge zone the total number of bacterial cells (active and inactive) ranges from 62 thousand to 1.2 million; in the zone of oil deposits the number ranges from 105 thousand to 90 million; and in the discharge zone, the range is from 42 thousand to 3.4 million.

These data show that a tremendous amount of organic material accumulates in ground water (up to 9 million cells per milliliter of water). The quantity of saprophytes decreases somewhat from the zone of recharge toward the discharge zone, an entirely natural relationship, since the most active decomposition of ready prepared organic material should occur in hydrogeologic zones nearest the zone's surface. However strange it may seem, the quantity of saprophytes decreases sharply in the vicinity of oil deposits as well, despite the fact that large quantities of ready prepared organic material are present here. The presence of high-temperature water (up to 83-89°C) does not put a stop to the life activity of microorganisms; it fosters the development of thermophilic races.

The distribution of physiological groups of microorganisms presents a picture no less interesting. All three of the previously indicated zones contain bacteria that destroy cellulose; the bacteria enter at the zone of recharge. The number of these bacteria decreases sharply from the zone of recharge to the zone of discharge (Fig. 5). In the vicinity of oil deposits and at the zone of discharge, the bacteria decompose cellulose at high temperatures (from 55 to 89°C). Oxygenizing bacteria are also found in all three zones. They decrease somewhat in numbers toward the vicinity of oil deposits. Heptane-oxidizing bacteria are also encountered in all three zones. They are most abundant about oil deposits and are least abundant at the zone of discharge. Methane-oxidizing bacteria are present in small numbers near oil deposits and at the zone of discharge. Methane-forming bacteria in a carbonic acid environment, with hydrogen present, may be found in all three zones in approximately equal numbers. Thermophilic methane-forming bacteria appear in the zone about oil deposits. Methane-forming bacteria that grow in ethyl alcohol are found only at the zone of recharge. Desulfurizing bacteria have been identified in all three zones. They are most abundant in the vicinity of oil deposits and at the zone of discharge.

Thermophilic desulfurizing bacteria are also found in the vicinity of oil deposits and at the zone of discharge. Desulfurizing bacteria, apparently developing at the expense of amino acids introduced from the earth's surface, are also present in all three zones in approximately equal numbers. Thermophilic denitrifying forms have been discovered about oil deposits and in the zone of discharge. Thionic acid bacteria are found only in the vicinity of oil deposits and in the zone of discharge, i.e., where sulfide ions are present.

All this indicates that ground water in all zones of its movement contains large numbers of various microflora, including thermophilic races, which give rise to very complex biochemical processes. The food supply for subsurface microflora is obviously found in stable organic remains, chiefly of land plants. Most important of these are cellulose and amino acids, the latter being decomposition products of albumins. Some types of bacteria also generate gases, such as methane, heptane, hydrogen, and hydrogen sulfide; others require these gases (heptane-oxidizing, oxygenizing, and methane-oxidizing), a fact pointing to the presence of corresponding biocoenoses in ground water.

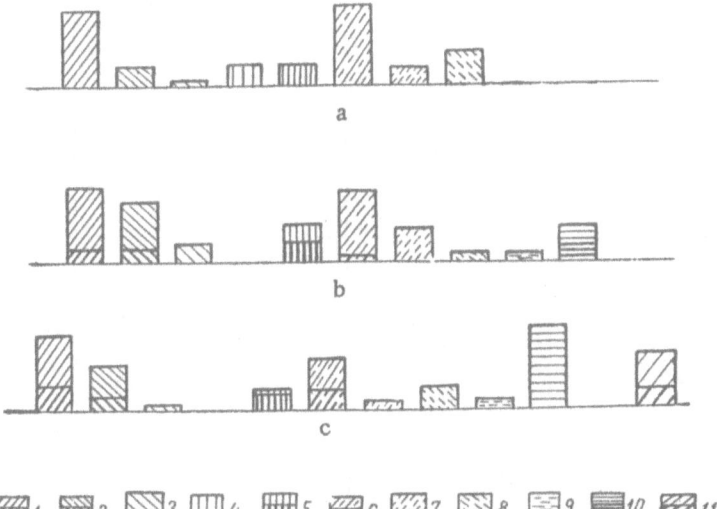

Fig. 5. The average abundance of bacteria in water, according to zones. a) Zone of recharge, b) zone surrounding oil deposits, c) zone of discharge. 1) Denitrifying; 2) desulfurizing; 3) methane-forming in an environment with calcium acetate; 4) methane-forming with ethyl alcohol; 5) methane-forming with carbonic acid and hydrogen; 6) cellulose-destroying; 7) heptane-oxidizing; 8) oxygenizing; 9) methane-oxidizing; 10) thionic; 11) denitrifying in Liske medium. The heaviness of the lines signifies intensity of bacterial development at temperatures between 55 and 89°C.

The quantitative relations of individual species of bacteria attests to the fact that such material as cellulose and amino acids penetrate, with ground water, to great distances and great depths in the sedimentary-rock layer. Methane is apparently produced in all three zones by the activity of bacteria growing in various environments. Methane-oxidizing bacteria appear only in the deeper zones, obviously where methane-forming bacteria will have produced sufficient quantities of the gas.

Preliminary results of microbiological investigations, using data only from 1955, for waters in the Chokrak and Karagan deposits in the Dagestan region are shown in Table 12. The abundance of individual physiological groups of microbes is indicated in this table on a five-point scale (data from Z. I. Kuznetsova).

The data in this table clearly show that the development and dissemination of microflora are practically the same in the waters of the Karagan and Chokrak strata as in the Grozny oil district.

However, the waters in the Khadum strata and in Lower Cretaceous formations, with which the gas deposits of Dagestan are associated, are essentially different from those in the vicinity of oil deposits. First, the number of bacteria in general, especially of saprophytes, is much smaller in the vicinity of gas deposits. Secondly, the waters in the strata in such areas do not contain desulfurizing, putrefying, and methane-forming bacteria in an environment with calcium acetate. Thirdly, comparatively small numbers of denitrifying bacteria are found in Liske medium, in addition to thionic acid and, in part, heptane-oxidizing forms. And, fourthly and lastly, such bacteria as denitrifying species in a medium with citric acid and as cellulose-destroying and methane-forming species in an environment with hydrogen and carbonic acid are found in approximately equal numbers.

TABLE 12

Microflora in Ground Water in the Dagestan Region

Sample locality	Computed quantities		Average abundance of bacteria by physiological groups, on a five-point scale									
	total number, thousands per ml	Saprophytes, no. of colonies per ml	Denitrifying forms		desulfurizing	Thionic	Putrefying	Cell-destroying	Methane-forming			Oxidizing
			medium with citric acid	Liske medium					$H_2 + CO_2$	Calcium acetate	H_2	Heptane
Zone of recharge	788 11	5200 142	4	1	1	0	4	4	3	1	1	1
Deposits	844 14	600 2	2	1	2	3	1	3	2	1	1	2
Zone of discharge	1152 19	2626 6	2	2	3	4	2	4	3	0	1	1
Ground water in the vicinity of gas deposits (Khadum horizon and Lower Cretaceous)	132 7	46 4	3	1	0	1	0	3	3	0	1	1

We should add to this the fact that the investigations of G. A. Mogilevskii and Z. I. Kuznetsova, extending over many years, indicate that methane-oxidizing and heptane-oxidizing bacteria are widespread in ground water, even close to the earth's surface. For example, species of such bacteria were identified by them in springs on the Stavropol upland, in Neogene aquifers in the steppe region of the Crimea, and in springs along the Oka-Tsna ridge.

Our preliminary investigations in the Grozny region and in the Dagestan ASSR generally show that water in aquifers in which oil and gas deposits occur contain various stable remains of land plants, of which cellulose (for example) is decomposed continuously throughout the entire aquifer (from the zone of recharge to the zone of discharge). The kind and numbers of microflora in ground water associated with oil and gas deposits exhibit several distinctive characteristics.

In considering all the results of the individual investigations on questions concerning environments and the possible means of converting separate organic residues to oil, the following statements may be made.

1. The conversion of organic residues to oil must take place at moderate temperatures (on the order of 125-130° and lower) and at a comparatively high pressure under conditions favoring reactions with decrease in volume in an alkaline and, chiefly, reducing environment.

2. The initial stage in the conversion of organic residues involves fundamentally the loss of oxygen (according to G. L. Stadnikov and to N. A. Orlov, and V. A. Uspenskii), either by loss of H_2O and CO_2 (according to A. F. Dobryanskii) or by separation of carboxyls and hydroxyls (according to G. L. Stadnikov); this may entail the formation of brown high-molecular substances with condensation of the aromatic series.

3. Most investigators believe that hydrogen enrichment is effected by hydrogenation. The source of free hydrogen may be found in the decomposition of water molecules by alpha radiation, in microbiological processes (hydrogen and butyric acid fermentation), and in hydrogen sulfide (according to N. F. Balukhovskii). The view of G. L. Stadnikov and L. V. Khmelevskaya that free hydrogen originated at depth is improbable.

A. F. Dobryanskii has proposed that hydrogen is redistributed by temperature effects, catalytic processes, and radioactive elements.

4. The source material for oil represents the final products of decay: organic acids (saturated and unsaturated fatty acids and aromatic acids), including naphthenic acids, amino acids, anhydrides of acids, alcohols, ketones, acetaldehydes, terpenes, humic material, and metal-organic compounds.

5. The majority of investigators hold to the view that there is a genetic relationship between the individual constituents of petroleum and the predominance of certain end products of decay of organic material. Thus, nitrogenous compounds are associated by all investigators with albumins, proteins, and amino acids; aromatic hydrocarbons and tarry substances with unsaturated acids, humic material, and terpenes; naphthenic hydrocarbons with naphthenic acids, unsaturated acids, tars, and terpenes; methane hydrocarbons with waxes, esters, anhydrides of acids, acids in general and, in particular, with saturated acids, alcohols, ketones, and acetaldehyde. Sulfur compounds are associated with H_2S introduced from outside. A. F. Dobryanskii accepts the view that there is a possible transition from aromatic hydrocarbons to naphthenic hydrocarbons, and then to methane hydrocarbons.

6. Radioactive processes promote the formation of oil by decomposing water to oxygen and hydrogen and also facilitate the formation of combustible gases (experiments of Zheslen), or they may produce liquid hydrocarbons directly.

7. Microbiological processes also, above all, foster the formation of oil by forming gases of biochemical origin (H_2, H_2S, NH_3, CH_4, and CO_2), and, possibly, they lead to the formation of bituminous substances and unsaturated naphthenic hydrocarbons (according to T. L. Ginzburg-Karagicheva).

8. Catalytic processes promote the processes of oil formation when particles of clay or any of various microcomponents are present.

9. Some investigators believe that the processes of oil formation are facilitated by the presence of chlorine, iodine, bromine (N. T. Shabarova), and ferrous oxide in saline water (G. I. Teodorovich).

To this we might add that oxygen compounds probably represent incompletely used residues of the initial source material (various organic acids and several other compounds).

All the conditions, both in relation to the physical environment (the presence of final decay products of organic material) and in relation to possible mechanisms of converting the decay products, indicated by the various investigators and discussed in the above summary are actually found in ground water.

Let us enumerate them:

a) the temperature of ground water increases gradually from the mean annual value of a given locality to a value of the order of 150°;

b) the pressure also increases gradually from atmospheric to hundreds of atmospheres;

c) in ground water the environment is chiefly alkaline and reducing, except for the upper zone of fresh water (where oxidation may occur slowly);

d) free hydrogen has been identified, having formed by alpha radiation and by microbiological processes; free H_2S, hydrous sulfides, various end products of decomposition of organic substances, clay particles in sandy aquifers, diverse microflora, haloids (chlorine, bromine, and iodine), and ferrous oxide have also been found.

Before going on to a discussion of the processes leading to the formation of gases and petroleum, let us pause briefly to consider further the composition of petroleum.

As is well known, natural petroleum consists of a complex mixture of gaseous, liquid, and colloidally dissolved solid substances. The quantity of dissolved gases in places amounts to several hundred cubic meters per ton of petroleum.

Colloidally dissolved substances (tars, paraffins, and other constituents) make up as much as 40% of some oils.

The hydrocarbon composition of natural oils ranges between wide limits. For example, A. A. Kartsev [1954] has stated that the extreme values for methane hydrocarbon content range from 0 to 75%, for naphthenic hydrocarbons from 20 to 80%, and for aromatic hydrocarbons from 5 to 60%. From this it follows that the constituent of natural petroleum most consistently present is the group of naphthenic hydrocarbons.

The hydrocarbons of natural petroleum that constitute its main bulk contain admixtures of oxygen, nitrogen, and sulfur compounds in small quantities.

Oxygen compounds found in petroleum include the following: saturated-type acids ($C_nH_{2n-2}O_2$), and a few aromatic acids and phenols. The acid content ranges from 0 to 2%, in places reaching as much as 3% (naphthenic medicinal oil).

Sulfur in oils occurs in the form of colloidal solution of elemental sulfur (up to 1% of the oil) and in the form of naphthenic sulfides (thionaphthene). Most of the sulfur occurs in tars, i.e., in the colloidally dissolved part. There are low-sulfur oils ($S < 0.3\%$), medium-sulfur ($5 > 1\%$), and high-sulfur oils ($S > 4\%$), such as the Chusovoi, Termez, and Mexican oils.

Nitrogen, like sulfur, is found chiefly in tars, i.e., again in the colloidally dissolved part of petroleum. Nitrogenous compounds found in petroleum include porphyrin, pyridine, and quinoline.

The tars apparently consists of naphthenic-aromatic groups containing atoms of O, S, and N and having molecular weights from 300 to 1000.

Asphaltenes are distinguished from tars by a lower solubility. They are generally present in amounts less than 1%. A relatively high content of asphaltenes is sometimes associated with a high tar content in petroleum, but, at other times, it is not. A high content of asphaltenes is sometimes observed in high-paraffin oils.

Of the microcomponents in petroleum ash, the most interesting from a genetic point of view are vanadium and nickel. The first sometimes constitutes as much as 30% of the ash (Ishimbai), and the second may form up to 12% of the ash. According to investigations of L. A. Gulyaeva and I. I. Romm, vanadium and nickel are associated with tars and occur in them in some kind of complex compound.

In considering the composition of natural petroleum, we should emphasize above all that petroleum is a mixture of substances of varying physical composition.

Since the formation and migration of gases, liquids, colloids, and solid substances obey essentially different laws, resulting from the physical nature of these substances, there can scarcely be any grounds for stating that all components of petroleum form uniformly in some particular place in the shell of sedimentary strata; i.e., it can hardly be thought that, in nature, small drops or any other minute particle of diffuse-disseminated natural oil form and are then collected in oil deposits.

If we take into consideration the laws of formation and migration of gaseous, liquid, colloidal, and dissolved solid substances, then we must suppose that the components of natural oil that are different physically are most likely formed in environments which, though possessing some traits favorable for the formation of oil, are not equally favorable in all parts of the formation or at all times during the oil-forming process. Apparently, the most probable course in nature is for disseminated components of oil, different in their physical and chemical relations, to form, and then, migrating according to the laws characteristic of their properties during favorable geologic and hydrogeologic circumstances, they collect in minute and large oil deposits. This is the first conclusion, and it has fundamental significance in understanding the origin of petroleum.

A second important conclusion, arising from a consideration of the marked percentage fluctuation in the individual components of petroleum, is that one may hardly conceive that only physical peculiarities of their migration can lead to qualitative differences in the composition of oils at various oil fields and oil deposits. If we begin with the most likely view, that the individual components of natural petroleum are associated with different initial substances (and it is possible that everyone now agrees with this position), then we have no grounds for maintaining that the initial substances accumulate under natural conditions in any strictly definite percentage relations that obtain in the now-observed average composition of natural oils. It is most likely that the initial substances accumulate in various quantitative relations, depending primarily on peculiarities in the composition of the plant community and on the conditions of the decomposition and accumulation of the organic material.

In addition, the formation of petroleum by means of hydrogenation unconditionally requires the presence of free hydrogen in order for the decomposing plant remains to become enriched in this element, i.e., in order for hydrogenation to occur. The quantity of hydrogen forming in nature, of course, is not quite sufficient to effect fully the process of hydrogenation or to give it a definite trend. Since hydrogen is rarely found in natural gases, it should be stated that it is formed in quantities insufficient for converting all the initial source material to hydrocarbons of the petroleum series. Thus, the quantity and, in part, the composition of natural petroleum are limited and regulated by the quantity of hydrogen participating in the process of hydrogenation. And, finally, since all the necessary substances for the reactions involved in the formation of natural-oil components come together in accidental concentrations, it is then obviously, as well, that the occurrence of these reactions is limited by the minimum concentrations of the substances. Therefore, after the reactions that lead to the formation of hydrocarbons, there may remain some quantity of the initial substances, in various stages of transformation, as unused residues, which may become impurities in the resulting petroleum.

In general, it may be stated that the quantity of natural-oil components that form and their group and individual composition depend on: a) the composition and relative concentrations of the initial substances, b) the quantity of free hydrogen, c) the conditions favorable to the course of several long-term specific processes, above all the processes of hydrogen, carbon dioxide, and methane fermentation, and d) favorable or unfavorable migration and accumulation jointly of the disseminated components of natural petroleum, which formed in different places and at different times.

All this brings to pass extensive variations in the quality of natural petroleum.

In seeking a solution to the problem of the origin of oil, it is necessary to consider, of course, the individual composition of substances constituting the various natural mixtures combined by us in the general concept of natural petroleum.

Dry and rich gases are distinguished among the gases accompanying an oil deposit. The first contains a large percent of methane and a considerable quantity of carbon dioxide. Rich gases contain no carbon dioxide; they are characterized by a small percentage of methane and by a relatively high content of its gaseous homologues. For example, according to S. S. Nametkin [1955], the dry gases in Dagestan contain up to 93-95% methane, up to 30-40% carbon dioxide, no more than 2-3% ethane, 1.5-2% propane, and tenths of a percent of butane and isobutane. The rich gases of the Grozny region contain up to 10-50% methane, no carbon dioxide, up to 7-17% ethane, up to 12-35% propane, up to 6-11% butane, up to 6-11% isobutane, and up to 5-11% gases having the formula C_5H_{12} and having high molecular weights.

Hydrocarbons of the methane series present in petroleum consist chiefly of pentane and hexane. For example, according to S. S. Nametkin [1955], pentane and benzene distillates constitute up to 40% of the total; hexane forms 40% in distillates at 68-70°C, up to 25% in distillates at 70-72°C.

The heptane content is small. Other liquid homologues of methane are present in very small quantities.

Solid homologues of methane from $C_{16}H_{34}$ to $C_{60}H_{122}$ (paraffins) are present in quantities from 0.1-2 to 9.9-11.2% (Karachukhar, Zykh). In this connection we distinguish between low-paraffin and paraffin oils.

In Table 13 we show the physical properties of the hydrocarbons in the methane series that are pertinent to this study.

From the data shown in Table 13 it is clear that the methane hydrocarbons up to and including isobutane, which boils at a temperature of −11°C, may be found only as free gas or as molecular dissolved gas. The next three hydrocarbons, with boiling points from 28 to 69°C, being weakly soluble in water, may be found in either a liquid or gaseous state, considering the high temperatures of water at depth; and they ever tend to separate completely from the water when there is any decrease in formational pressure. The boiling point of heptane is near that for water, and it may therefore migrate with the water for considerable distances. Octane and nonane will be found in the liquid state, separating from water because of their negligible solubility.

We should recall here that the content of these hydrocarbons in petroleum is very small. And, lastly, paraffins, having melting points ranging from 19 to 70°C, will be found in the liquid state, separating from water when the temperature is lowered.

Unsaturated hydrocarbons of the fatty series are almost nonexistent in petroleum. They have been identified reliably only in Canadian oil.

TABLE 13

Physical Properties of Hydrocarbons in the Methane Series

Name of hydrocarbon	Physical state at 100°C	Formula	Melting point,°C	Boiling point,°C	Sp. gr. temp. °C	Solubility of gas in water (in 100 g of water at a temp. of 18-20°C)
Methane	Gas	CH_4	—186	—164	$\frac{0.415}{-164°}$	9 ml
Ethane		C_2H_6	—172	—93	$\frac{0.446}{0°}$	4.7 cm³
Propane		C_3H_8	—	—44.5	$\frac{0.536}{0°}$	6.5 cm³
Butane (n)		C_4H_{10}	—135	+1	$\frac{0.600}{0°}$	Insoluble
Isobutane		C_4H_{10}	—145	—11	—	13 cm³
Pentane (n)		C_5H_{12}	—131	+36.3	$\frac{0.648}{0°}$	0.036
Isopentane		C_5H_{12}	—158	+28	$\frac{0.639}{0°}$	Insoluble
Hexane		C_6H_{14}	—94	+69	$\frac{0.677}{0°}$	0.0138
Heptane		C_7H_{16}	—97	+98.3	$\frac{0.701}{0°}$	0.0052
Octane	Liquid up to $C_{60}H_{122}$	C_8H_{18}	—56	+125.8	$\frac{0.719}{0°}$	0.0015
Nonane		C_9H_{20}	—51	+150.5	$\frac{0.733}{0°}$	Insoluble
Hexadecane		$C_{16}H_{34}$	+19	+287	$\frac{0.771}{10°}$	—
Eicosane		$C_{20}H_{42}$	+37	+205	$\frac{0.778}{37°}$	—

Naphthenic hydrocarbons, having a cyclic structure and being similar to the hydrocarbons of the methane series in their chemical properties, are frequently called cycloparaffins. Monocyclic naphthenes have been found in petroleum: cyclopentane, cyclohexane, and cycloheptane; constituting the main bulk of the petroleum. The physical properties of these compounds are listed in Table 14.

From the data in Table 14 it follows that the bulk of the naphthene hydrocarbons are not soluble in water and have boiling points slightly below or slightly above water; not mixing with water, they may migrate with it to considerable distances and may separate in the gaseous state when pressures become lower.

The physical properties of the aromatic hydrocarbons (benzene and its homologues, and condensed aromatic systems — naphthalene) are listed in Table 15.

From the data in Table 15 it may seem that the aromatic hydrocarbons are predominantly liquids in hot waters at depth.

The various tarry and asphaltic substances (mineral oils, oil tars, asphaltenes, asphaltic acids, colloidal asphaltenes) dissolve readily in hydrocarbons, forming true solutions with them. Of the sulfur compounds hydrogen sulfide and mercaptan (R — SH) are rarely present in petroleum; thiophenes (cyclic compounds) have been reported in only individual localities. Elemental sulfur and aliphatic sulfides are found most frequently in petroleum. All sulfur compounds are insoluble in water.

TABLE 14

Physical Properties of Naphthenes

Cycloparaffins or naphthenes	Physical state at 100°	Formula	Boiling point,°C	Specific gravity	Solubility in water
Cyclopentane	Gas	C_5H_{10}	50.5	0.7506	Insoluble
Cyclohexane	"	C_6H_{12}	72	0.7474	
Cycloheptane	Liquid	C_7H_{14}	118	0.8108	—

Nitrogenous compounds found in petroleum include ammonia, simple amines, pyridine, and quinoline. The lower amines dissolve easily in water. For example, at 12°C water will dissolve 1150 times its volume of methylamine. Pyridine* has a specific gravity of 0.98, a melting point of 42°C, a boiling point of 115.4°C, and will dissolve without limit in water. Quinoline** has a specific gravity of 1.09, a melting point below 15°, and a boiling point of 237.7°C. Nitrogeneous compounds are thus present either as gases or liquids.

TABLE 15

Physical Properties of Aromatic Hydrocarbons

Aromatic hydrocarbons	Physical state at 100°	Formula	Boiling point,°C	Melting point, °C	Specific gravity	Solubility of gas in water (per 100 g of water at 22°C)
Benzene	Gas	C_6H_6	80.4	5.5	0.879	0.1856
Toluene	Liquid	$CH_3C_6H_5$	110.6	−95	0.866	0.0492
Xylene	"	$(CH_3)_2C_6H_4$	144	−47 to +13.2	0.876	0.0130
Cumene	"	$C_3H_7C_6H_5$	153	—	0.866	—
Naphthalene*	"	$C_{10}H_8$	218	80.1	—	—

* The melting point of naphthalene is 80°C.

Oxygen compounds in petroleum include saturated acids with the composition $C_6H_{2n-2}O_2$ and naphthenic acids $C_nH_{2n}CO_2$, which have high boiling points (on the order of 215-275°C). Salts of alkali metals, in which oxygen is present, dissolve very readily in water. The specific gravity of the acids is near unity (0.950-0.988). Naphthenic acids and their salts may migrate extensively with ground water. Phenol, identified in petroleum, has a specific gravity of 1.07, a melting point of 181.2°C, and it dissolves in water in any quantity at high temperatures.

In considering the brief information cited above concerning the composition of natural petroleum, concerning the general views on the origin of oil, and concerning the results of our investigations in the Grozny-Dagestan gas and oil region relative to the formation of individual components in natural petroleum, we may make the following analysis.

GASEOUS COMPONENTS

In studying the gaseous components of petroleum we should first of all solve the problem relative to the formation of methane and carbon dioxide, since these two gases are the principal "dry" gases, and one of the methane series constitutes the bulk of the "rich" gases.

* Heterocyclic compound with a single atom of nitrogen in the ring.
** Condensed molecule of a heterocyclic compound consisting of a benzene nucleus and a pyridine nucleus.

It is well known to everyone that methane forms by biochemical means in anaerobic environments (swamps, peat bogs, flood plains, etc.) near the earth's surface. It may also be formed in the same way in ground water where anaerobic conditions exist below the boundary beyond which atmospheric oxygen does not penetrate.

Our investigations in the Grozny region (data of Z. I. Kuznetsova) have revealed that ground water contains bacteria that destroy cellulose in zones of recharge (Chernye Mountains), in the vicinity of oil deposits, and at sites of discharge (emergence of thermal springs). The number of these bacteria decreases sharply down the dip of an aquifer, i.e., from zone of recharge toward zone of discharge. In the zone of discharge and in the vicinity of oil deposits are found bacteria that decompose cellulose at high temperatures (from 55 to 89°C).

Methane-forming bacteria, growing in an environment of carbon dioxide and hydrogen, occur in all three of the indicated zones in approximately equal numbers.

The investigations of V. Omelyanskii show that the decomposition of cellulose by Bacillus cellulosae methanicus is accompanied by the generation of considerable quantities of CH_4, methane fermentation occurring at first, then hydrogen fermentation. There are special groups of bacteria that decompose cellulose at temperatures of 35-51°C and even at temperatures of 60-65°C. During methane fermentation gases are liberated in amounts equal approximately to 50% of the quantity of decomposing cellulose; of this, CO_2 constitutes about one-third of the total gas formed, CH_4 about two-thirds. There are data that indicate methane fermentation in a pure form does not exist. Methane is obtained after carbon dioxide forms according to the scheme

$$CH_3COOH \rightarrow CH_4 + CO_2$$

$$CO_2 + 8H \text{ (or } 4H_2O) \rightarrow CH_4 + H_2O$$

During prolonged activity of cellulose-destroying bacteria, only methane is obtained.

Hydrogen fermentation also occurs during the decomposition of cellulose. According to our investigations (data of Z. I. Kuznetsova), hydrogen-fermenting bacteria are found in all three of the above-indicated zones in the ground water of the Grozny region (recharge zone, vicinity of oil deposits, and discharge zone). Their number decreases somewhat near oil deposits. Hydrogen fermentation of cellulose yields CO_2 and H_2 in addition to organic acids. Both gases are liberated in approximately equal amounts, their volume being about one-third that of the decomposing cellulose.

Thus, the presence of cellulose and the proper bacteria lead to the formation of carbon dioxide, methane, and hydrogen; in the process approximately from one-third to one-half of the decomposing cellulose changes to gas. The longer the process is active, the more methane is formed. We think the latter fact is very important, since, hydrogeologically, the faster ground water moves the more carbon dioxide tends to become the dominant gas, but the slower the movement of ground water the more the process shifts to the formation of methane.

The formation of such gases as methane, carbon dioxide, and hydrogen by biochemical means in aquifers occurs, naturally, throughout the aquifer and during the entire time that decomposed plant remains (especially cellulose) are present in the ground water.

Some of the gas that forms migrates down the dip of the formation in a molecular-dissolved state in proportion to its solubility, the temperature of the ground water, and the pressure. The excess gas that has formed is disseminated, partly in the free state, but, in part, migrating upward along the dip of a formation, it reaches a gas-oil trap, forming a pure gas deposit if the trap is not subsequently invaded by the liquid components of petroleum; or it forms a gas cap on an oil deposit. Gas caps are later supplemented by gases separating from the liquid components of the oil.

At present, we have no direct information of ethane, propane, and butane. However, we have no grounds for denying a dominant role for microbiological processes in the formation of these gases.

To this we should add that the formation of carbon dioxide, hydrogen, and methane takes place in ground water in an anaerobic environment, not only by decomposition of cellulose, but also by decomposition of various organic acids, including amino acids. For example, formic acid is decomposed by Bacillus formicum to carbon dioxide and hydrogen; the biochemical decomposition of acetic and butyric acids also generates methane and carbon dixiode in addition to other products.

In general microbiological processes produce the following effects:

a) further, more profound decomposition of organic material occurring in ground water, chiefly in the direction of forming various organic acids and such gases as methane, carbon dioxide, and hydrogen;

b) direct formation of the principal constituents of natural gas (methane and carbon dioxide) and the chief gaseous component of oil (methane);

c) preparatory work in developing an environment favorable for the formation of the other components of oil, by generating, first, hydrogen and then hydrogen sulfide in ground water.

It is possible that some part of disseminated hydrocarbons is formed in the bodies of microorganisms.

The role of microbes in transforming nitrogenous and sulfurous substances will be discussed below.

LIQUID HYDROCARBON COMPONENTS

The problem of the formation of liquid hydrocarbon components of natural petroleum is very complex. However, thanks to the investigations of A. F. Dobryanskii, G. L. Stadnikov, V. A. Uspenskii, O. A. Radchenko, V. N. Ipat'ev, T. L. Ginzburg-Karagicheva, N. T. Shabarova, and many others, the field of possible source materials and reactions leading to the formation of hydrocarbons has been greatly restricted in comparison to what was known before these studies.

The principal source material for the formation of hydrocarbons is most likely represented by organic acids and their salts (saturated and unsaturated fatty acids, naphthenic acids, and aromatic acids). Wax is an important factor in the formation of large quantities of methane hydrocarbons.

Humic material is associated with the formation of tar and asphaltenes. Terpenes, of course, have fundamental significance as source material. Metallic-organic compounds may also have some value in the generation of hydrocarbons.

Almost all investigators ascribe great significance to hydrogenation and catalysis in the processes of carbon dioxide formation. This is quite natural, since the chemical composition of oil is distinguished from the initial organic material chiefly by enrichment of the organic substances in hydrogen, and catalysis lowers the required energy for reactions involving the combining of hydrogen.

Hydrogenation, above all other processes, needs free hydrogen, the presence of which in ground water is not yet clearly understood, since it is known that free hydrogen is rarely encountered in natural gases. However, it should be noted that free hydrogen has been found at a number of places in amounts ranging from 0.4 to 66.5% of the total volume of gases. For example, 66.5% was measured in Nizhnii Tagil, 27.3% in Stavropol, 4.6-12.7% in the Southern Urals, and other considerable amounts elsewhere.

Free hydrogen is formed chiefly through microbiological processes and, in part, through the decomposition of water by alpha radiation.

Actually, according to C. E. ZoBell [1947], each gram of organic material furnishes, during decomposition, 40-50 cm^3 of hydrogen, but one gram of carbohydrates yields up to 80 cm^3 of this gas. The same investigator gives a list of 29 microbes that produce H_2 during decomposition of glucose, cellulose, and many other organic substances. It is still necessary to keep in mind that hydrogen is expended in the process of denitrification, in reducing sulfates to H_2S (each gram of hydrogen sulfide uses at least 2600 cm^3 of H_2, and each cubic centimeter of CH_4 requires no less than 4 cm^3 of H_2), and, lastly, is easily oxidized by oxygenizing bacteria. If we keep in mind the expenditure of hydrogen going into the hydrogenation of organic material, then we should be surprised, not that it rarely escapes to the earth's surface, but that it is everywhere present in natural gases and, sometimes, even in relatively large quantities.

Radioactivity represents another source of hydrogen in ground water. It is well known that water molecules break up into hydrogen and oxygen when exposed to alpha radiation. The oxygen is used in various oxidizing processes. In addition to large quantities of H_2, according to Breug, Koenig, Sherrel, and others, hydrocarbons are also formed by alpha radiation. The formation of hydrogen by alpha radiation takes place locally, of course, i.e., only where radioactive elements are present in the ground water.

Probably the amount of H_2 formed by any or all the processes is a limiting factor for associated processes, including the hydrogenation of organic material; i.e., the amount of hydrocarbons that forms is directly dependent on the amount of H_2 that forms.

The absence of hydrogen gas in ground water means the absence of liquid and colloidal components of petroleum.

The formation particularly of the lower methane hydrocarbons may take place by decarboxylization of carbonic acid according to the scheme

$$R - COOH \rightarrow R - H + CO_2,$$

i.e., by separation of CO_2 from the salts of carbonic acid. As is well known, this reaction may occur at comparatively low temperatures, but only when aided by catalysts, including biological agents. Other processes, of course, may also take place.

For example, such a process is the conversion of dicarbonic acids through hydrogenation of ketones to naphthenes according to the scheme

Adipic acid   Cyclic ketone   Naphthene-cyclopentane

In addition, saturated hydrocarbons with open chains (paraffins) and with six or more atoms of carbon may change to cycloparaffins even without the addition of supplementary energy (i.e., they may change to naphthenes), and, with the addition of energy, link carbon chains may form from propane, butane, and pentane. On the other hand, during catalytic hydrogenation at temperatures of 120° (cyclopropane) and 180° (cyclobutane), cyclic hydrocarbons may be broken down and butane and propane formed from them. However, cyclopentane is hydrogenated only at temperatures above 300°, and cyclohexane, in general, does not hydrogenate at all. Thus, methane hydrocarbons may change to napthene hydrocarbons. The reverse process is rather improbable. However, this does not exclude the possibility that propane, butane, and even pentane may form from corresponding cycloparaffins.

In general the possibility of a process taking place and the specific importance the process may have may be determined if we will make a sufficiently complete study of the composition of the organic material in aquifers associated with oil and gas deposits.

In the literature on the formation of oil, it is noted with complete justification that high pressures and temperatures promote the formation of hydrocarbons. We think, though, that other factors should also be considered, but in complete agreement with their actual value in physicochemical processes. For example, one might speak of the minimum temperature necessary for hydrogenation and catalysis to occur, or of the maximum temperature at which hydrogenation gives way to destructive hydrogenation. Between these limiting temperatures, changing values lead only to acceleration or retardation of a reaction.

At present we know nothing of minimum temperatures. A maximum temperature probably exists since, according to V. N. Ipat'ev's experiments, a smooth process of temperature and pressure increase is always disturbed when it is accompanied by the dissociation of the source material. For example, the decomposition of products that formed by hydrogenation of naphthalene occurs at 400°.

Pressure, as is well known, primarily fosters the formation of substances with less volume and generally leads to acceleration of a reaction, but sometimes it slows reactions, as was noted above for the decomposition of alcogels.

Favorable factors for the formation of petroleum that should always be noted are the alkalinity and salinity of the water, the presence of iron, and a reducing environment.

Alkalinity is favorable for the neutralization of organic acids by alkali salts and alkaline-earth metals. The presence of alkaline waters in the vicinity of ore deposits is probably associated, as noted above, with the formation of soda during hydrogenation and catalysis.

The salinity of water, according to our meager experience, promotes the coagulation of material in ground water.

Iron compounds, like aluminum compounds, may be catalysts. And, lastly, a reducing environment is required for anaerobic processes, since it is in an aerobic environment that organic material, in the final analysis (not considering their mineral components), is decomposed to carbon dioxide and water.

There are other ways of collecting hydrocarbons in ground water. C. E. ZoBell [1947], for instance, has noted that some hydrocarbons and other components of oil are synthesized in the cells of plants, animals, and bacteria and are preserved intact.

Jankovski and ZoBell [1944] have shown that desulfurizing bacteria are able to form hydrocarbons from fatty acids, the hydrocarbons accumulating in bacterial cells and possibly separating out after the death of the organism.

M. V. Fedorov [1955] has also stated that amino acids of the fatty series may be reduced to hydrocarbons in an anaerobic environment according to the following equation:

$$R - CH \cdot NH_2 \cdot COOH + 2H = R - CH_3 + CO_2 + NH_3.$$

And, lastly, many disseminated hydrocarbons are formed on the earth's surface. Hydrocarbons of the paraffin series accumulate in the waxes of higher plants, and cyclic terpenes occur in needles and pitch of conifers (see Table 3).

As we have already mentioned, the recent investigations of P. V. Smit on the content of hydrocarbons in Recent sediments have shown that hydrocarbons are found not only in Recent marine muds and sands, but also in the deposits of deltas, swamps, shallow muddy lakes, rivers, forests, peat bogs, and soils.

The conclusion that disseminated hydrocarbons are widespread regionally in deposits of various origins is perhaps the most interesting and important of all the investigations of P. V. Smit.

If there are disseminated hydrocarbons in the atmosphere, in the soil, in peat bogs, in swamps, in lacustrine, and fluviatile deposits, in land plants, and in the bodies of microorganisms, then, being borne by infiltering water into the shell of sedimentary strata, they may certainly accumulate at first in the upper water-bearing layers and then, being carried downward by the flow of ground water, they may be carried to the depths of the sedimentary shell.

All these means of forming and accumulating hydrocarbons in subsurface waters do not exclude, but rather supplement, each other. However, the specific value of each is not entirely clear at the present time.

NITROGEN COMPONENTS

As is well known, the decomposition of albumin substances in an aerobic environment leads to the formation of ammonia. This process is called ammonification. It has been ascertained experimentally that bacteria consume two grams of nitrogen out of 100 g of decomposing organic material (or out of 50 g of carbon). Therefore, if the ratio of carbon to nitrogen is 25:1, all nitrogen is completely consumed by bacteria. The addition of carbohydrates somewhat decreases the amount of ammonia being formed.

Later, the quantity of ammonia in ground water is regulated by two processes: nitrification ($NH_3 \rightarrow N_2O_3 \rightarrow \rightarrow N_2O_5$) and denitrification ($N_2O_5 \rightarrow N_2O_3 \rightarrow NH_3$).

As we know from hydrogeology, the first of these two processes is dominante in upper waters; i.e., the more NH_3 and N_2O_3, the fresher we consider the impurities in ground water. On the other hand, a small quantity of N_2O_3 and a large quantity of N_2O_5 attests to protracted processes of contamination. Both these processes may take place at the same time, but, in waters at depth, denitrification is possibly the dominant one. For example, according to our investigations (data of Z. I. Kuznetsova), the ground water in the Grozny region was found to contain denitrifying bacteria in approximately equal numbers in zones of recharge, in the vicinity of oil deposits, and at zones of ground-water discharge. These data indicate that the admixture of ammonia in petroleum is probably related to processes of denitrification.

Simple amines are also products of decomposition of organic material that contains nitrogen.

Amino acids of the fatty series are formed most abundantly during decomposition of albumin substances. in an anaerobic environment. The fatty series may be reduced to hydrocarbons according to the following equation:

$$R - CH \cdot NH_2 \cdot COOH + 2H = R - CH_3 + CO_2 + NH_3.$$

If sulfur is also present in the amino acids, H_2S is obtained, and, in addition, methyl mercaptan and mercaptans of other types.

Amino acids of the aromatic series, when decomposed in an anaerobic environment, yield a series of great interest from the viewpoint of formation of petroleum products. For example, phenylalanine (an amino acid of the aromatic series) yields benzoic acid and tyrosine-phenol; tryptophane gives skatole and indole. The latter compound belongs to the group of growth substances (indole-acetic acid, known under the term heteroacetate). It has now been ascertained that decahydronaphthalene petroleum processes growth properties.

Thus, the decay of albumin substances in an anaerobic environment not only may be responsible for nitrogenous admixtures in oil (phenol, mercaptan, growth substances) but may lead as well to the formation of some of the hydrocarbons.

SULFUR COMPONENTS

As may be seen from what has gone before, various kinds of mercaptans may be formed during the decomposition of amino acids containing sulfur. Elemental sulfur is probably formed in associated with the life activities of desulfurizing bacteria (which produce hydrogen sulfide) and of sulfur bacteria, the sulfur compound being later oxidized to elemental sulfur and being deposited within the cells. Sulfur bacteria, as is well known, have long been known to occur in waters associated with oil deposits. For example, the widely known purple bacteria, which give the water a rose color, have been found in such an environment. Our investigations in the ground water of the Grozny region have shown that desulfurizing bacteria occur in small quantities in recharge zones; they are considerably more abundant in discharge zones and even more abundant yet in the vicinity of oil deposits.

We should turn our attention to the fact that the ground water in the Grozny region contains thiobacteria, but they are absent entirely at recharge zones. They are present in the vicinity of oil deposits and are even more abundant at discharge zones, i.e., in the water from known hot hydrogen sulfide springs. Thiobacteria oxidize sulfur, thio compounds ($N_2S_2O_3$), tetrathionic compounds ($N_3S_3O_6$), and H_2S to sulfuric acid without depositing sulfur within the cells. Thus, sulfurous admixtures and their abundance in oil are probably associated with microbiological processes.

OXYGEN COMPONENTS

Of the oxygen admixtures in petroleum, naphthenic acids attract attention primarily. It has previously been thought that these compounds are the products of oxidation or of biochemical alteration of hydrocarbons in the marginal zones of oil deposits. As the preceding data have shown, from the viewpoint of the hypothesis developed in this book, naphthenic acids represent one of the fundamental source materials for the formation of hydrocarbons. Admixtures of naphthenic acids in petroleum are unused residues from the reactions that lead to the formation of hydrocarbons.

The validity of this point of view is rather convincingly strengthened by the results of our investigations, which show that naphthenic acids are found in fresh-water springs in the Chernye Mountains of the Grozny region, a fact that fully supports the view that there is a stream of naphthenic acids moving from recharge zones of water-bearing horizons toward oil deposits. From the hydrogeologic point of view, it is entirely probable that the products of oxidation or of biochemical alteration in the marginal parts of oil deposits, such as in the Sunzha Range, were carried down to the depths of the Sunzha syncline, after which they migrated upward along the dip of the formations till they emerged at the surface in the Chernye Mountains.

The presence of phenol in petroleum may be explained by either of two processes. In the first, phenol may be formed by microbiological decomposition of amino acids of the aromatic series (tyrosine); in the second, phenol may come from the effect of highly concentrated chlorine on organic material. The latter process is rather well known from experience with water pipes, in which strongly chlorinated drinking water containing organic material is given a phenol odor.

Asphaltenes, carbenes, asphaltic acids, and asphaltenic colloids most likely represent unused (and to some degree altered) residues chiefly of humic material, which have migrated in solution in hydrocarbons, in colloidal form, or in coagulated clots (forming because of the coagulating effect of saline waters).

All these substances, together with the oils and petroleum tars from all the components of natural petroleum, have low mobility. During migration, they lag behind other components, entering the oil deposit last, and imparting a high specific gravity to the oil in the marginal zones. Some of these substances are completely separated from the gaseous and liquid components of the petroleum, tending to fill various types of cavities in the aquifers. When they accumulate in large masses, which should be a characteristic occurrence during formation of source material from temperate-climate plants rich in tars, and when pressures are high and fracturing due to tectonic movements extensive in the host rocks, they may inject large cavities and fractures in considerable quantities, forming various types of asphaltic deposits.

We shall not pause here to discuss the microelements present in petroleum ash, since this matter was touched on above.

We shall note merely that some microelements (vanadium, nickel, and iron) are associated with porphyrins, which are widely distributed in petroleum, in petroleum derivatives, in oil shales, and, to a lesser degree, in coals. According to their structure, porphyrins belong chiefly to the chlorophyll group. The investigations of O. A. Radchenko and L. S. Sheshina [1955] show that sulfur-poor oils (less than 0.7%) are free of porphyrins or contain them in insignificant quantities. The quantity of porphyrins in sulfur-rich oils is 10-1000 times greater.

In our view, porphyrins are primary and are stable residues of chlorophyll. Sulfurous oils enriched in vanadium and, in part, nickel complexes are possibly associated with the decomposition of ferments (biocatalysts) of albumin substances.

In concluding the present chapter we should say something about the formation of deposits of ozocerite and asphalt.

As is well known, ozocerite (or black wax) is a transparent oily liquid at 58-62°, having a specific gravity of 0.8-0.9 and consisting of high-molecular hydrocarbons (paraffins and ceresins).

Asphalt, which softens at temperatures of 70-110° and which has a specific gravity ranging from 1 to 1.2, consists of nearly nonvolatile hydrocarbons of the naphthene series with admixtures of oxygen and sulfur compounds.

Ozocerites and some asphalts might be examined from the viewpoint of the hypothesis developed in this book: first, as methane oils rich in paraffins and, secondly, as naphthene-aromatic oils with a considerable admixture of oxygen and sulfur compounds. An enrichment in paraffins indicates extensive introduction of wax-like compounds into the ground water (waxes, cutins, and other integumental elements of plants protecting them from excessive evaporation), which may occur in warm, moist climates, i.e., in tropical and subtropical regions.

Components of oil rich in paraffins and ceresins fall into zones of lower temperature when migrating during the development of an oil deposit; because of this, high-molecular paraffins and ceresins are separated from the total migrating mass, forming separate deposits or else accumulations associated with petroleum. Under these conditions, as is well known, paraffins are extracted from petroleum by artificial means, i.e., by mechanical filter pressing or by preliminary chilling of the petroleum mixture. According to I. M. Gubkin [1937] ozocerite veins are actually petroleum intrusions and, in appearance, they suggest volcanic intrusions.

When large quantities of tarry substances accumulate with the source material, which may take place either in tropical regions or in regions of conifer forests in a temperate climate, asphaltic deposits may be formed (in a similar manner).

It is also possible that adsorption phenomena and coagulation, in addition to decreased temperatures, may be important factors in the formation of such deposits. Of course, these statements do not include deposits of asphalt formed by oxidation of petroleum.

In all discussions on the origin of petroleum thus far outlined by us, it has been assumed that the source material in aquifers was introduced from the surface of the earth. But there are other possibilities. For one, source material may have been introduced into ground water, not from the earth's surface, but from sedimentary rocks, chiefly, of course, marine rocks. For another, ground water may have become enriched in this source material by infiltration of water along water-bearing horizons that also contain some quantity of organic material of marine origin. These views also should be tested by further study, although they seem at the moment to be less probable than the accepted one.

Organic material in pelitic rocks undergoes profound changes during diagenesis, chiefly of the coal series (this does not exclude the rare discoveries of liquid bitumens in such rocks), and physicochemically is firmly bound to the mineral part of the rock, forming adsorbed organic substances, cement in the rock, and detritus. It is therefore difficult to extract from the rock, even by organic solvents, and it is almost completely insoluble in water. The transfer of organic material from marine pelitic rocks to water-oil reservoir rocks is possible only in the earliest stages of diagenesis, directly following the mud stage. But, in this case, the source material will be introduced into the water-bearing horizons again in the form of organic acids and other end products of decay, and this will be simply a supplementary source of furnishing the aquifer with organic material. Organic substances that are present in water-bearing rocks may enter the ground water. But, since their solubility is very slight and since the water-bearing horizon itself was formed first, passing through all stages of diagenesis, and then the ground water accumulated in it, this method of enriching ground water is, if effective at all, most likely of relatively little value.

All that has been said about possible processes taking place in ground water and leading to the formation of oil represents only a rough scheme or first approximation, needing to be subjected further, as far as studies of organic material and microflora in ground water are concerned, to fundamental tests to prove its validity. But, as an initial statement to guide extensive scientific research in this field, one may hope that it will be sufficiently acceptable and fruitful for more profound study of the mechanism of formation of the material constituting natural petroleum.

Chapter IV.

THE MECHANISM OF OIL-DEPOSIT FORMATION

This chapter does not present a thorough examination of the problems of oil migration and the formation of oil deposits. It contains some remarks on hydrochemical indicators for petroleum prospecting, introduced in this book in order to show that if we adopt the proper view (that oil is formed in water-bearing horizons) then we may, just from this point of view, examine the questions of oil migration and the mechanism of forming oil deposits.

These are complex and, at the same time, unnecessarily confused problems. The confusion is due to the fact that, if we hold to the point of view that oil is formed in marine clay rocks, it is then necessary to explain the very first stage of oil migration, i.e., the transfer of the oil from impermeable (to oil and water) clay rocks into rocks that are permeable to oil and water. In general, it becomes necessary to solve a rather contradictory problem. In the meantime, one may observe in nature, anywhere and at any time, the inverse processes in layers forming the sedimentary shell: porous and fractured rocks being filled with water and gas at recharge zones and, possessing pressures different from overlying slightly permeable beds, supplying water to these strata. An aquifer may acquire water from neighboring clay rocks only if the aquifer is drained artificially or if zones of lesser pressure are produced in it. From the hydromechanical point of view it is impossible to conceive of clay rocks, filled with some particular liquid, yielding this liquid to overlying sandy rocks which are, of course, already filled with water since the pressure difference necessary for this process is much too small in nature to overcome the resistance of clays impermeable to oil and water. Compression of the clays in the given example solves nothing either, since one cannot conceive how, in nature, tectonic forces can compress clay strata without doing the same to overlying sandy beds. Naturally, not only will the clay beds and the overlying sandy strata be compressed, but the whole sequence of formations in the section will be affected. Confusion regarding the question of oil migration and the mechanism of forming oil deposits is also found in those views in which the source beds are thought to occur at considerable distances from or great distances below the oil deposit. In these views full range is allowed the "migration fantasy," as a result of which the essential point simply gets out of perspective. All these complex and confused questions become simple and comparatively clear if we agree on the view that oil is formed in the aquifers. When this is done, the migration of water, oil, and gas may be explained by strict hydromechanical laws and computations. This does not mean, of course, that our task becomes extremely simple, since the joint movements of water, oil, and gas through a porous environment represent, in themselves, very complex and very difficult problems both in respect to the mathematical formulation of any solution, and in respect to the definition of initial and terminal conditions, dictated by the geologic-hydrogeologic parameters, which alter from the moment oil forms to the time an oil deposit is produced. Despite this, however, in considering the focus of oil formation to be transferred from clay rocks to aquifers, the problem of oil migration is automatically moved from the region of general and extremely indefinite considerations to the region of physical and mathematical computation.

As is well known, the problem of oil migration and of the formation of oil deposits has occupied a large place in the literature. Some investigators, alluding either to observations made at individual oil deposits or to special laboratory experiments performed by them, attempt to ascribe to some force or other a chief or universal significance. Various theories have arisen in respect to this: gravitational, hydraulic, gravitational-hydraulic, filtrational, diastrophic, sedimentational compaction, and capillary. But it is improper to seek to devise a single scheme for the migration of oil, governed by some sort of universal force. There is no doubt that, during the migration of oil, a number of physical phenomena appear that simultaneously affect the migration. But, some of these become dominant under conditions of oil formation that approach natural conditions for them. If therefore seems to us that the mechanism by which oil deposits are formed should be considered, not in general, but in agreement with concrete types of oil or hydrogeologic traps. In the future these considerations may be refined and be applied to actual oil deposits.

It was pointed out above that all oil traps are actually hydrogeologic traps and possess, as a rule, one common property: a marked change in permeability (to oil and water) in the zone of the oil deposit. From this it follows that, first, oil deposits are formed by joint seepage of water, oil, and gas, and, secondly, in regards to time, a hydrogeologic trap forms first and the oil deposit is then formed in it.

Finally, we may consider the problem of the physical state in which oil migrates, since there are numerous and very diverse opinions on this matter. Different investigators have suggested that oil may migrate in the following way:

1) in an aqueous solution as gas, oil, or some intermediate substance;

2) as vapor, capable of being converted to oil or gas by changes in temperature and pressure;

3) as a gas, capable of changing to oil, by means of catalytic processes, after arriving at the reservoir;

4) as crude petroleum, containing or not containing gases;

5) as viscous asphalt-like primary petroleum;

6) as unsaturated hydrocarbons or some other highly mobile products of the primary conversion of organic material;

7) as droplets of oil or as gas bubbles, the small dimensions of which do not prevent their passing in water through the pores of the rock;

8) as oil droplets and gas bubbles, the relatively large size of which leads to their alteration as a result of the Jamin effect;

9) as a film in an aqueous environment;

10) as oil and gas, moving in a continuous mass.

Some of these views do not contradict but rather supplement one another. For example, the ideas proposed in points 7 and 8 are but refinements of the first point in regard to possible dimensions of the oil drops. If we assume that oil migrates extensively, and this is supported by all the data obtained in productive operations and in geologic exploration, then the views expressed in points 5 and 10 seem quite improbable, since oil cannot migrate far in such a state (continuous or asphalt-like masses). If the oil occurred in such form, it should either occupy the site where it formed or should be found very near that site. To propose the movement of oil entirely in a gaseous form (points 2 and 3) is to suggest something we find no basis for, since it requires that we acknowledge that, along the course of migration, the pressure and temperature must be above the critical values for all the constituent parts of the oil. The concept of oil movement as a film is also unacceptable because films form only on a free surface of water. There can be no such free surface in pressurized subsurface water.

In general, in seeking a solution to the problem of the physical state of migrating oil, it is necessary to keep in mind the following: first, that the accumulation of the initial source mateial, the transformation of this material, and the subsequent migration of oil occurred in an aqueous environment, continuously, however slowly, moving (i.e., in an environment of ground-water flow) and, secondly, that both the source material and the individual constituent parts of the oil must exist in a physical state proper for them at the temperature and pressures obtaining in each actual geothermodynamic zone that subsequently cuts across an aquifer. Some data on critical temperatures and pressures are given in Table 16.

Finally, a third point is that oil forms at separate points in this aqueous environment, not in a compact mass. From this, it follows that the gaseous components of oil migrate chiefly in a dissolved state, in part as very small free bubbles of gas, whereas the liquid components migrate in a dispersed state in aqueous emulsions (water − oil) and, in part, in a dissolved state if we consider the slight solubility of hydrocarbons in ground water. We may be convinced of this by the practice of extracting oil from flooded wells (water drive) into which aqueous emulsions actually flow; these emulsions show various stages of saturation with oil. As will be seen from discussions

TABLE 16

Critical Temperatures and Pressures for Several Materials
(after V. A. Sokolov)

Substance	Critical temperature, °C	Critical pressure, atm
Water	365	—
Carbon dioxide gas	32	—
Nitrogen	− 147	33.5
Methane	− 82	50.0
Ethane	32	45.2
Propane	96	45.0
Butane	153	36.0
Pentane	197	30.0
Hexane	234	26.8
Heptane	267	—

below, the break up of emulsions into their constituent parts, or the separation of oil from water, may take place in different types of hydrogeologic traps because of various causes.

In general, the liquid and gaseous components of oil that have formed in various parts of the aquifer occur in a disseminated state, developing because of different degrees of solubility and because of variations in specific gravities (water, oil, and gas); the unstable emulsion constantly tends to separate and gradually does so as external stimuli are added to strengthen the internal forces tending to cause the emulsion to separate.

After these preliminary remarks, let us go on to examine the mechanism by which oil deposits are formed (in respect to the various types of oil traps).

During long intervals of geologic time, oil that has formed in a sequence of beds that dip uniformly and have slowly become buried may migrate with the ground water, which will flow with some velocity, and may gradually accumulate in deep segments of the water-bearing strata. When the ground water moves very slowly — almost always the situation when drainage is absent in the aquifer or when the beds dip very gently — the oil and the gases dissolved in it may separate from the moving water because of differences in specific gravity, or they may even migrate upward and, encountering an impermeable layer or a facies change in the formation up dip, form oil deposits (of the Maikop type) without the participation of any other stimuli.

One of the most important factors leading to the formation of oil deposits in domal structures is tectonic folding. During tectonic movements various types of secondary folds are produced, some elongated, some doubly plunging. The tectonic stresses increase the previously existing hydrostatic pressure and create hydrodynamic conditions characterized by variable pressures, not only along the vertical but also along the horizontal. At such times the zones of highest pressure will coincide with the limbs of the folds, and the zones of low pressure will coincide with the newly formed positive structures and, to a lesser degree with negative structures (the crests and troughs of folds, respectively).

As is well known from geology, such a distribution of pressures during disharmonic folding produces flow in plastic beds (such as clays) toward the crests of folds. The same feature, only in incomparably greater degree, is found in ground water, which will move with increased velocity, together with the water-oil emulsion, at times in one direction, at other times in all directions toward domal uplifts.

The same thing may be said about the distribution of temperature conditions; i.e., positive and, to a considerable lesser degree, negative structures will exhibit comparatively low temperatures, whereas the limbs of folds will show comparatively high temperatures. An increase in temperature will be associated with the tremendous compression of the rocks in synclinal zones and on the limbs of folds. A decrease in temperature in domal structures will be accentuated because of nearness of the earth's surface.

This distribution of temperature conditions should also lead to the separation of water and oil, a fact supported by the experiments of Emmons[1921-1922]. His experiments have shown that when oil, water, and gas seep through curved tubes and when the temperature is simultaneously increased, the constituents quickly separate, the oil and gas accumulating in the crests, i.e., in the bent segments of the tubes. Under such conditions, the emulsions of oil and water will generally move with increased velocity from zones of high pressure and temperature to zones of low pressure and temperature, i.e., from the limbs to the crests. A decrease in temperature and pressure as well as variations in specific gravities will promote the separation of oil-water emulsions.

In addition, in accordance with Bernoulli's law, at each bent segment of a stratum the velocity of flow will be somewhat increased, and, because of this, there will be a supplementary lowering of the pressure directly in the water-bearing horizon, similar to that encountered in water pipes. From experience with water pipes, it is well known that,where a convex bend occurs in a pipe,gas tends to be expelled, but the gas does not force itself through the water; it rather accumulates in such quantities that it forms an air lock, completely stopping the flow of water in the pipe. Because of this, all convex sections of pipe must be equipped with special devices, escape valves, which automatically permit accumulating gases to escape.

The separation of water-oil-gas emulsions is greatly facilitated by the shaking of the beds, during folding movements. In this relation it is of interest to note two circumstances. A. F. Dobryanskii has pointed out that when material moistened with oil is shaken, the oil may be easily extracted from an emulsion. And, in addition, it is well known that increased yield at gas wells by the act of exploding torpedoes in them is not due entirely to the formation of additional fractures; but the gas begins to separate in large quantities because the explosion produces an elastic agitation in the rock and in its contained mixture of water and gas. An instructive example is found in the vicinity of Derbent, where gas is strongly emitted after earthquakes.

Some very interesting experiments were performed by Wyckoff and Botset on the joint filtration of water and gas of water and oil. The chief results of these experiments are shown in Figures 6 and 7. An examination of these graphs shows that, on filtering emulsions consisting of 80% water, the permeability of sand falls to 45% relative to the value for filtering water alone; but, the permeability for gas and oil drops to 4-5% because of stopping the pores with oil and gas. When emulsions contain only 50% water, the permeability for water drops to 8%, and,when the water content is approximately 20%, the permeability for water approaches zero, but, on the other hand, the permeability for oil and gas increases up to 90%; i.e., under these conditions water sticks in the pores, but gas and oil move almost as well as if water were completely absent. This means that shaking and decrease in pressure lead to the separation of gases and oil from emulsions, and the proportion of these constituents in the water gradually decreases. When the content of oil and gas in an emulsion is about 20%, these constituents filter freely into the upper part of anticlinal arches, gradually forming ever-enlarging caps of gas and oil.

Depending on the quantity of gas or oil that has separated (either one or the other) and the thickness of the formation, the anticlinal arch may become stopped up, and the movement of all three components of the mixture will cease in the vicinity of the oil deposit. The stopping of pores by gas is fully confirmed, as well, in the practice of flooding wells (water drive) in the oil fields of Tataria. A decrease in pressure in exploitation wells leads to the separation of gases,which stop pores in the strata and thus halt the movement of oil from flooded wells; this process considerably decreases the effect of flooding.

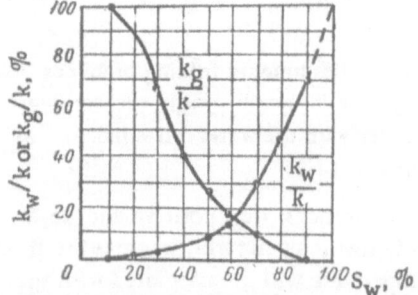

Fig. 6. Graph showing the joint filtration of water and gas.

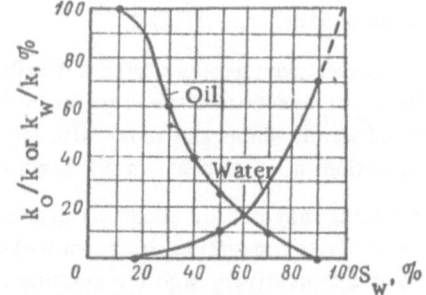

Fig. 7. Graph showing the joint filtration of water and oil.

In seeking a solution to the problem of oil migration, it is necessary to keep in mind that, according to the investigations of Beecher and Parkhurst [1926], the viscosity and surface tension of oil are markedly lowered by gases dissolved in it. Thus, at a pressure of 35 km/cm^2 and a temperature of 21.6°C, gas dissolves in oil in such quantities that the viscosity is lowered 50%. At a depth of approximately 1300 m, the viscosity of oil may be reduced to the value of kerosene.

In general, during the formation of oil deposits in the crests of anticlines, several factors are in effect sequentially and, in part, simultaneously: diastrophic, hydraulic, filtration, and gravitational.

The above-described mechanism of forming anticlinal oil deposits has been confirmed by several known observations. For example, the hanging deposits of the Apsheron Peninsula (displaced deposits on the limbs of the folds) have water-oil contacts inclined at angles of 30° [S. I. Mironov, 1952]; such a phenomenon may be the result of hydrodynamic conditions, but not of hydrostatic. The magnitude of the slope of a water-oil contact will depend on the rate of movement of the emulsion. When the rate of flow of water and water-oil emulsions is low, the slope may have an inclination opposite to the direction of ground-water flow or it may be horizontal. When the rate of flow is high, on the contrary, the inclination will be in the general direction of ground-water flow; and, lastly, when the rate is exceptionally high (such as in coarse gravel in the arch of an anticline), it is theoretically possible that no oil deposit will form in the structural closure or that oil will accumulate only in small quantities. Such huge accumulations of oil as found in the Sabine and Bend anticlines and the Seminole uplift can hardly be explained, as noted by Ries, without assuming that migration has occurred on a grand scale from extensive oil-collecting areas.

Deposits in the Applachian region, in Pennsylvania and West Virginia, are situated in the outer, western zone of the fold system, strictly zoned, or parallel to the range. The greatest oil production is found in the north, the amount of oil decreasing sharply to the south; i.e., folds near the oil source accumulate large deposits, but folds at greater distances retain smaller quantities.

But, very interesting and concrete data have been furnished by S. P. Maksimov [1954] on oil deposits in Carboniferous and Devonian rocks at Samarskaya Luka. He has noted the following rules of occurrence for this region.

1. The composition and specific gravity of the oil change from east to west, the oil becoming heavier (for example, the Devonian oil increases in specific gravity from 0.8033-0.8202 on the east to 0.8580-0.8798 on the west, and aktsizny tars (tar-like substances obtained by a sulfuric acid test) and asphaltenes increase from 3.31 to 11.5%; a similar picture is observed for Carboniferous oils.

2. The entire sequence of sedimentary strata shows a regional rise in this same direction, from east to west.

3. The oil has migrated chiefly from east to west, and, possibly, also to the north, i.e., away from the depressed limb.

4. The migration occurred after the Zhigulevsk disturbance.

5. The oil in the Devonian beds and that in the Carboniferous strata of the middle Volga region formed independently of each other.

The mechanism of forming shielded deposits is considerably simpler. Except for the relationship to the geologic nature of the shield (tectonic-faulting in friable formations, stratigraphic-unconformities and lithic changes into clay zones, salt deposits, and other facies along the dip of sandy beds), shielded deposits, after they form, always stop the movement of ground water in the segments of the bed and give rise to conditions necessary for the age-long separation process in the emulsions — water, oil, and gas. This process is greatly aided by agitation of the strata, because of elastic waves, each time there is an earthquake, the agitation being a decisive factor for sharp separation of oil and gas from the emulsions; and gravitation is also significant (because of different specific gravities of the components constituting the emulsion) during slow separation of the components.

The mechanism of forming massive deposits may be variable. For example, massive deposits in structural traps (Fig. 8), i.e., where hydrodynamic distribution of pressures may obtain during formation of the deposit because of deformation processes, the mechanism may be to a considerable degree like that suggested for the development of deposits within strata in anticlinal arches. For massive deposits in erosional traps (erosional highs surrounded by rocks impermeable to oil, including chemically precipitated sediments – Fig. 9), the gravitational factor may be the principal one in the process of separation, amplified by agitation. And, lastly, massive deposits in more permeable zones of dense rocks (Fig. 10) most likely form in the manner described below for deposits bounded by slightly permeable rocks, keeping in mind that commerical quantities of oil accumulate in limestones possessing zones of secondary porosity due to leaching along ancient weathering surfaces.

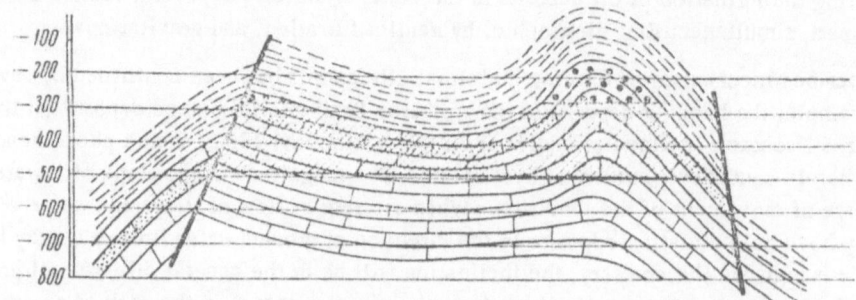

Fig. 8. Section of a massive deposit in a tectonic (structural) trap (after I. O. Brod).

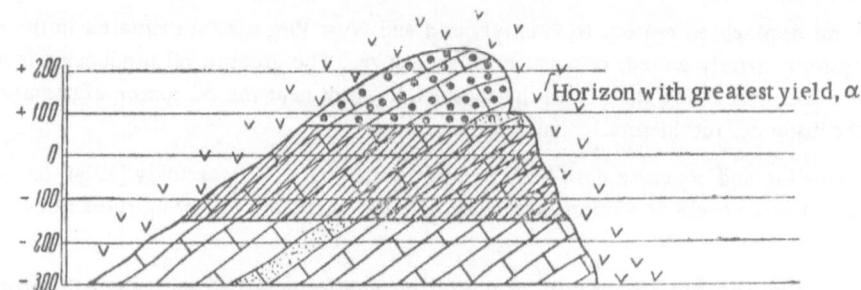

Fig. 9. Section of a massive deposit in an erosional trap (after I. O. Brod).

Figure 11 shows an occurrence of oil in a coarse variety (gravel) of water-bearing and sandy strata; and Figure 12 shows a deposit in sandy lenses of low-permeability rocks (clay rocks). In all three types of oil occurrence, emulsions, filtering through relatively impermeable rocks, reach individual sites in lenses of highly permeable rock or find zones of increased filtration. This indicates that the transfer of emulsions from slightly permeable to highly permeable rocks is entirely possible; but the reverse process is not, since part or all of the petroleum sticks in the highly permeable lenses, converting these lenses to zones impermeable to water. To this it should be added that one cannot, of course, state that highly permeable lenses are dry (free of water) before the oil arrives. It is assumed here that the effect of capillary forces is scarcely possible, since such forces always tend to produce movement toward fine-grained material; i.e., in all three cases capillary forces should hinder the entrance of the oil and should favor its escape from the lenses, especially (it should be kept in mind) since the surface tension of oil is approximately one-third that of water. However, this process does not occur, and, consequently, capillary forces have little to do with the formation of this type of deposit. In addition, a number of investigators have recently shown that capillary forces may be effective from an interface of more or less permeable rock only for a very short distance: some tens of centimeters (or, more precisely, about 58 cm, according to experimental data). Nevertheless, the presence of capillary forces in streams of filtering liquid is doubtful since these forces are associated with the formation of menisci, which appear at the boundary between free gas (atmosphere) and the liquid. In a moving stream of ground water there is no such interface, and there are no menisci; consequently, the existence of capillary forces in this flow of gravitational liquid is doubtful.

To explain the mechanism of formation of such a deposit, Illing performed some very interesting experiment in 1933 and 1939 [1948]. According to these experiments, when a mixture of 10% oil and 90% water filters

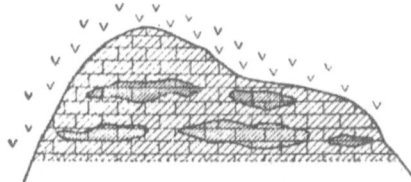

Fig. 10. Section of a massive deposit in a reef zone (after I. O. Brod).

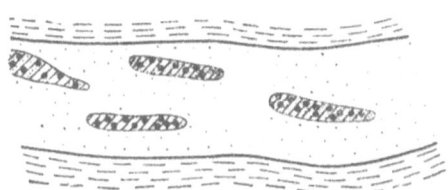

Fig. 11. Section of a deposit surrounded on all sides by water (after I. O. Brod).

through alternating layers of coarse- and fine-grained sands, the water passes completely through all the layers of sand, but the oil without exception goes only through the coarse-grained sand. It passes only into the upper part of the fine-grained sands. It follows then that oil mixed with water freely passes from fine-grained sands to coarse-grained sands. The reverse passage occurs only with difficulty. There are thus grounds for stating that oil deposits, lithically bounded on all sides, are generally formed by filtration.

We should also mention something concerning oil deposits in synclinal structures. Such deposits are rarely found in nature, but they nevertheless do occur. The mechanism of their formation may be as described in the following. Oil, as well as highly mineralized waters, moving along the dip of a formation, should naturally accumulate (and actually does) in the lowest synclinal segments of the aquifer; i.e., such deposits might be considered normal accumulations of oil, developing without the participation of deformation processes. But, the mechanism of forming synclinal deposits may also be somewhat different if we consider that the maximum compression during the formation of folds occurs on the limbs, and, therefore, the syncline should represent a zone of somewhat lower pressure (like an inverted anticlinal arch). Therefore, some of the oil may be squeezed, not only into the positive arch, but, in rare cases, also into negative arch structures (synclines).

In concluding this chapter we should mention the term "water-free" deposits of oil and gas, commonly encountered in the literature on petroleum. Such a term always signifies only the circumstance that there is no well-defined oil-water contact in the well, or as established by drilling. Thus, I. O. Brod assigns such a term to deposits that occur in reservoirs bounded by lithic changes on all sides and saturated with gas and oil. A plane water-oil contact is absent at such a deposit; the contact is curved, enclosing the deposit on all sides. The existence of water-free gas deposits (i.e., deposits with no underlying water) is entirely possible, since ground water may be present at the time the gas deposit formed, and then, because of any of a number of causes, might break away from the lower plane of the gas deposit.

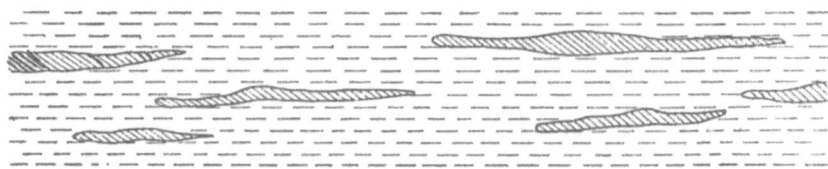

Fig. 12. Section of a deposit bounded on all sides by low-permeability rocks (after I. O. Brod)

From all that has been said concerning the formation of oil in aquifers and concerning the mechanism of forming oil deposits, it follows that, first, oil and oil deposits are formed individually or independently of each other in the separate aquifers and, secondly, the processes are continuous and, when the proper conditions persist, may be in effect from the time the aquifer formed to the present day. This has been shown very convincingly in the already mentioned paper of Z. A. Tabasaranskii [1954]. Quoting from that paper: "Concerning the existence of a post-Maikopian phase of oil migration, evidence is found in the presence of deposits of light oil and of gas as a gas cap beneath an unconformity" (Fig. 13). According to Tabasaranskii the oil moved from north to south in the Paleogene deposits of the Il'sko-Kholmskii region, as indicated by the fact that in the northern deposits some of the horizons contain light oil, but in the southern deposits, heavy and asphaltic. In addition, a very distinctive oil deposit, a double deposit so to speak, has been observed; it is shown diagrammatically in Figure 14.

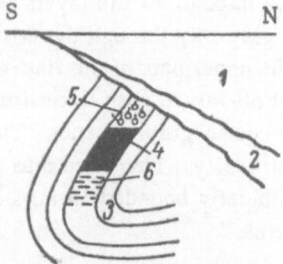

Fig. 13. Oil deposit beneath
an unconformity in the
Il'sko-Kholmskii region of
the northwestern Caucasus
(after Z. A. Tabasaranskii).
1) Miocene, 2) Maikopian, 3
3) Paleogene, 4) light oil,
5) gas cap, 6) water.

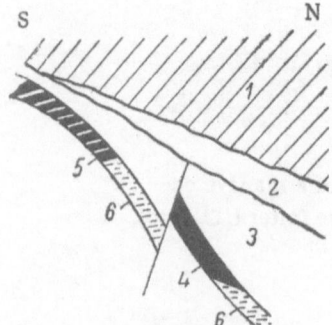

Fig. 14. Oil deposit in the
Il'sko-Kholmskii region of the
northwestern Caucasus (after
Z. A. Tabasaranskii). 1) Miocene, 2) Maikopian, 3) Kuma
series, 4) light paraffin oil,
5) heavy asphaltic oil, 6) water.

As a result, Z. A. Tabasaranskii, quite independently of the views we have expressed here, came to the conclusion that the described relationship apparently indicates that oil deposits develop over a long period of time and probably are still forming in the Il'sko-Kholmskii region; this view is supported by luminescent-bitumen investigations on particles of liquid hydrocarbons in the waters of the indicated region.

From the principle that oil deposits form independently, one interesting conclusion follows. It is this: as is well known, no definite or universal rules can be expressed in regard to changes in specific gravity and in composition of oils along their vertical distribution, a fact borne out by innumerable studies made on this problem. On the other hand, the search for rules representing changes in specific gravity and in composition of oils along water- and oil-bearing strata should be intensified, since this may yield several auxiliary indicators for petroleum prospecting, both relative to the mechanism of forming oil and gas deposits and relative to changes in quality of oils during filtration and treatment.

In this relation, we may allude to a number of papers that contain interesting and instructive conclusions. Among these are the papers of A. A. Kartsev, V. N. Kholodov [1954], M. V. Abramovich [1939-1941], D. V. Zhabrev [1939], A. A. Kartsev [1951], V. T. Malyshek [1940-1943], B. M. Sarkisyan [1947], and others.

But, the conclusions of S. P. Maksimov [1954] are of special interest. From his observations we find that, during migration from the site where the components of oil originate to the site where an oil deposit is formed, the gas components are held back, forming pure gas deposits; somewhat farther along deposits of light oil develop, with or without gas caps; and, lastly, even farther from the site where the components originated, deposits of heavy, asphaltic petroleum are formed. This relationship is illustrated in Figure 15, borrowed from the paper of S. P. Maksimov. The conclusions of Z. A. Tabasaranskii [1954] also seem to support this position.

In conclusion it is necessary to pause to consider two questions. For every new geologic hypothesis, it is completely natural that a question should arise concerning the effects of the hypothesis in practical geologic work, and, further, what better basis the new hypohtesis has when compared with previosuly evolved hypotheses. First, let us look at the first question.

To begin, the concept that the disseminated components of petroleum are formed within aquifers requires more profound study both of present-day hydrogeologic conditions and, especially, of paleohydrogeologic conditions. Inasmuch as the source material for the formation of oil has been assumed to come from decomposing remains of terrestrial plants, more intense study is necessary on paleogeography and paleobotany, with designations of the regional distribution of zones of plant associations, since this will to some degree determine the abundance of source material; and zonal peculiarities in the composition of plant communities lead to qualitative variations in the composition of the future oil. For example, plants of tropical countries, rich in wax, may supply the basis of chiefly paraffin oils, whereas tarry plants of temperature latitudes may give chiefly aromatic oils.

Fig. 15. Distribution of oil and gas deposits in a regionally inclined stratum (after S. P. Maksimov).

Since the direction of transport of the initial material will be predetermined by ground-water currents and the conditions of their accumulation will be controlled by tectonic structures, especially careful study of the geologic history of the region is necessary and, particularly, the history of tectonic movements and structural changes resulting from these movements.

Naturally, all these questions should be studied in combination with full knowledge of the geology of the particular region being considered.

All the above-indicated questions are being studied at the present time in the light of the requirements of existing, most widespread views concerning the origin of oil.

The difference is that all in the above-indicated group of questions must necessarily yield a clear picture of the paleohydrogeologic and paleobotanical conditions of a particular region. New concepts simultaneously and urgently demand a broad, regional study of the group and individual composition of organic material migrating in the ground water and an equally broad study of the microflora inhabiting the ground water. Such studies will permit us to gain a deeper understanding of the mechanism by which the individual components of natural petroleum are formed.

Both on the basis of existing views and from the viewpoint proposed in our hypothesis, geological exploration requires a study of all factors favorable to the formation of oil- and water-reservoir rocks, their reservoir properties, and the conditions favorable to the creation of oil and gas traps (associated with primarily tectonic folding movements, producing second-order structures). No less attention should be devoted to the study of possible development of other types of oil and gas traps. From this point of view, the new hypothesis introduces nothing new in the practice of geologic exploration; it merely once more emphasizes the immense practical importance of prospecting criteria relative to the identification of reservoir sites and the location of oil and gas traps.

Different occurrences of oil should serve, as always, as direct criteria in searching for oil deposits; only their practical value is unfortunately becoming ever smaller and smaller because of the search for oil deposits at great depths.

The possible depth of prospecting for oil deposits is practically unlimited from the viewpoint of the concepts developed in the present book. Theoretically, the limit may coincide with the depth to which ground water penetrates in the earth's crusts. According to this view, oil may accumulate in deposits in rocks of any origin (marine, continental, volcanic) if they possess suitable reservoir properties and if, individually or in combination, they show marked changes in filtration properties in zones where oil and gas traps may develop.

Perhaps, the greatest change in the new views concerning the origin of oil is found in the introduction of hydrochemical factors. A rather long time ago high salinity in ground water was considered an indirect criterion in searching for oil. More recently, a high content of iodine and bromium has been thought to be an indirect indicator of oil accumulation. From our point of view, these indicators may still be valid and may continue to be. But, they should be considered primarily as indicators of high concentrations of ground water that occupies the lowest position in a given structure; they should not be thought of as elements in any way directly associated with the formation of oil. High salinity in the water also creates a physiological barren zone and may put a stop to microbiological activity, giving rise to conditions leading to mass destruction of microorganisms and, consequently, to the accumulation of hydrocarbons that are found in the bodies of these organisms.

It has always been recognized that upper, fresh, flowing ground water is unfavorable for oil accumulation. As something of a general position, this is perhaps correct, since microbiological decomposition of organic remains continues to be effective in this ground-water zone. But, this decomposition has already begun to produce such gases as methane and carbon dioxide. Therefore, gas deposits may also form in flowing waters, and sometimes even oil deposits may develop. Consequently, the position stated above cannot be granted any universal significance.

69

Very frequently, attempts have been made to relate oil deposits to definite chemical types of water. As seen from all the preceding discussions, there is no strict relationship. Above all, ground water must furnish a reducing environment, but this environment exists in ground waters of the most diverse chemical compositions.

The sulfate content or lack of sulfate in ground water cannot supply a unique indication of oil content, since both types of water very frequently are found with no direct relationship to the processes of oil formation.

In recent years it has been proposed that a content of naphthenic acid that is high for a given region may serve as a direct index of oil accumulation; there is undoubtedly a genetic relationship between naphthenic acids and the formation of oil.

But this indication, from our point of view, has a double signficance. First, when the content of naphthenic acids is high, one may think that there may have been an abundance of source material but little hydrogen necessary for hydrogenation. In this case, oil deposits with small reserves may form. Secondly, the absence of naphthenic acids may signify that they have been completely changed to hydrocarbons, and cannot, therefore, attest to any absence of an oil deposit. However, on the whole, the presence of naphthenic acids in regions of recharge and adjacent regions is a favorable indicator for the occurrence of oil over a considerable territory.

In our opinion hydrochemical indicators of oil and criteria for prospecting and exploring for oil may be found in the presence of some particular organic material in the ground water. Such direct indicators may be organic carbon, organic nitrogen, the ratio of these two elements, the group and individual composition of several organic compounds, and, especially, the presence of naphthenic acids in districts near recharge zones.

Inasmuch as some microcomponents migrate as metal-organic compounds and as the residues of ferments and are genetically related to land plants, they also may be used as hydrochemical criteria in prospecting. Of the microcomponents, nickel and vanadium deserve first mention.

And, lastly, inasmuch as the microbiological processes in the formation of oil have tremendous significance, individual species of bacteria may be used as criteria in the search for oil. These bacteria include not only microbes that oxidize the various hydrocarbons (i.e., fixing them) but also microbles that produce conditions favorable for oil formation and the creation of gaseous hydrocarbons.

However, microbiological indicators should not be looked for in the upper zones of ground water, but along the dip of aquifers.

The development of new hydrochemical indicators for oil is, naturally, a very complex problem and requires the accumulation of a great fund of information concerning organic material and microflora in ground water; this task can be achieved only by the joint efforts of many investigators.

Let us now pass on to the problem concerning the relative merits of the basis for the proposed hypothesis.

It is impossible here, of course, to compare this hypothesis with the various and numerous views and ideas relative to the origin of oil. And, perhaps, there is no necessity to do this. It will be quite sufficient to make comparisons only with the most widespread opinions, and, of these, only those concerned with fundamental positions.

SOURCE MATERIAL

Most investigators accept the view that oil is of organic origin. We feel that this point of view is fully confirmed. The source material is generally taken to be the stable remains of decomposition of the animal and plant world, chiefly of marine basins, but partly from the continents. We assume that the great bulk of source material comes from the decomposition of terrestrial plants and that only blood, perhaps, of animals supplies an admixture of porphyrins with an iron complex.

The concept that the source material comes from the decomposing remains of marine fauna and flora has no real foundation. It rests on the acceptance of the following position: the birthplace of oil is a marine basin containing source beds, which possess some special, but unknown, properties and peculiar features leading to further conversion of organic material, not to compounds in the coal series (as is generally observed in nature) but to compounds in the oil or bitumen series. Thus, a logical conclusion serves as a basis for establishing an opinion concerning source material.

In contrast to this, the hypothesis proposed here includes the following factors that indicate a genetic relationship between composition of oil and composition of the material in terrestrial plants:

a) the group and, in part, individual composition of oil is best correlated with the group and individual composition of the substances obtained during decomposition of terrestrial plants;

b) the ratio of carbon isotopes in oil is very similar to that ratio in terrestrial plants, but is markedly different from the ratio in marine plants;

c) the group of nitrogenous compounds in oil and, in particular, pyridine and quinoline are also related most closely to terrestrial plants, since pyridine and quinoline are also found in coal;

d) arachic acid has been found in Galician oil;

e) vanadium and nickel occur in large quantities in petroleum ash.

Of course, all these arguments do not yet supply absolutely convincing proof, but the statements are all based on facts; they are not premature, approximate truths.

THE SITE OF OIL FORMATION

According to existing views, oil is formed in sediments in marine basins. This position rests on two facts. First, the presence of decomposing organic remains in marine water and the burial of these remains during sedimentation are undoubted geologic facts. Secondly, the great reserves of oil occur in strata of marine origin. This is also an undoubted geologic fact. However, it is also a geologic fact that decomposing remains of terrestrial plants accumulate in ground water.

Actually, according to the investigations of V. G. Datsko [1951], the content of organic carbon in the Black Sea is 3 mg/liter on the average. The distribution of organic carbon in a vertical direction is rather uniform, but the material does decrease somewhat with depth. According to V. G. Datsko, "the distribution and quantity of carbon below 500 m does not support the idea that organic material is accumulating in the deep waters of the Black Sea."

The presence of organic material in ground water has also been known for a long time, through large numbers of sanitary-hygienic analyses.

Our investigations on the quantitative determination of organic carbon in ground water in the Grozny region have shown that, in the region of recharge (Chernye Mountains), the content of carbon is 3.4 mg/liter on the average, i.e., approximately the same as that found in the Black Sea and in the zone of discharge (thermal springs of the Peredovye Ranges). In the vicinity of oil deposits, waters from high-yield wells contain 6.6 mg/liter of organic carbon; waters from low-yield wells contain 34.7 mg/liter. These data attest not only to the presence of organic carbon in ground water in quantities exceeding that in waters of the Black Sea, but also to the accumulation of organic carbon in a direction away from recharge zones toward the zones of oil deposits. From the hydrogeologic point of view, the accumulation of organic material in ground water cannot be doubted.

The statement that the greatest reserves of oil are found in marine sediments has but limited value, since oil is also present in continental beds and in volcanic rocks. On the other hand, the statement that both commercial and all noncommercial accumulations of oil are found in aquifers encounters none of the above-indicated contradictions; however, most oil is found in aquifers of marine origin since such strata are most favorable for developing aquifers of great extent.

THE MECHANISM OF OIL FORMATION

Strictly speaking, the mechanism of forming oil from source materials, either from the viewpoint of existing concepts or from the viewpoint of the hypothesis proposed in this book, is still not clear. And, according to all views, certain values have been assigned to the significance of microbiological processes, hydrogenation, and catalysis; all consider the effect of temperature, pressure, alkalinity of the water, reducing environments, etc.

In general, our position in relation to the mechanism of forming oil will be considered approximately the same as others. But, in transferring the site of oil formation from solid sediments to an aqueous environment, we may approach more concretely and in greater detail the question concerning the origin, not of the diffused and disseminated oil, but of its disseminated components, which have different physical states (gases, liquids, dissolved colloids, etc.).

THE INITIAL PHASE OF MIGRATION

In conceiving that oil is formed in source beds, we must inevitably devise some mechanism for transferring the oil from the source beds to the reservoir rocks. As seen from previous discussions, neither compression nor processes of reciprocal evaporation can, under actual natural conditions, be mechanisms of this type.

In our proposed hypothesis, such an initial phase of migration is generally unnecessary, since the oil is formed directly in the water and oil reservoirs.

In this connection, all the objections of the adherents of a theory of inorganic origin of oil essentially disappear.

FURTHER COURSE OF MIGRATION

In regard to this matter, everything remains fundamentally unchanged, except that the migrations of the individual components of oil are thought to have occurred separately in accordance with their physical properties and with the laws characteristic of each aggregate state of the substances constituting natural petroleum. Therefore, it is contended that oil is finally formed only in gas-oil traps.

In concluding this discussion, we wish to note that, although, in the light of the new hydrogeologic hypothesis concerning the origin of oil, the oil is thought to have formed in ground water rather than in hypothetical source beds, still this hypothesis, where possible, has retained all the best ideas on this subject furnished by our leading petroleum geologists and chemists specializing in petroleum. In particular, we have retained in their entirety the views concerning the organic origin of oil, the very extensive migration of oil, the continuous series of conversions of organic material, the anticlinal theory as a mechanism of forming oil deposits, the significance of types of oil traps, and, lastly, the possible physicochemical and biochemical processes that transform stable plant residues to petroleum.

We think that the hydrogeologic hypothesis of petroleum origin gives a better explanation and ties all the principal observations on oil deposits, made during geologic exploration and extraction operations, into a single, harmonious system. This hypothesis is readily open to verification, and, if it is confirmed, it promises definite results in the matter of developing valuable criteria for appraising oil occurrences and for oil prospecting. In general, we hope that, if not the entire hyopthesis as a whole, at least individual parts of it may be useful for further development of the science of petroleum.

Part II

PRELIMINARY RESULTS OF INVESTIGATIONS
IN THE GROZNY-DAGESTAN OIL REGION

Chapter V

A BRIEF SURVEY OF THE HYDROGEOLOGIC CONDITIONS
IN THE GROZNY-DAGESTAN REGION AND THE CONTENT
OF ORGANIC MATERIAL IN THE GROUND WATER

In view of the fact that the group and individual compositions of organic material in ground water are of signficance not only in seeking an understanding of the mechanism of oil and gas formation but also in developing new hydrochemical criteria for evaluating oil potential and techniques used in prospecting for oil and gas (based on the salt composition and the organic material in the ground water), it was decided, in order to obtain comparative data, to study the organic material and microflora both in oil regions and in regions known to be without oil.* The Grozny-Dagestan gas-oil region and the Moscow basin were selected for these studies. In the latter region, studies on organic material were made in fresh and slightly mineralized waters from Carboniferous beds, where numerous wells for water supply have indisputably shown that oil and inflammable gases are not present. Both regions were selected for our studies because there had been many years of previous investigations on the geology and hydrogeologic conditions in them. This made it possible, first, not to spend time and money on any auxiliary geologic-hydrogeologic work and, secondly, strictly to control the accumulation of geologic and hydrogeologic data when selecting samples of water for analytical work to determine organic material and to study microflora. And, thirdly, it furnished an objective scientific base for interpreting the results obtained.

The Grozny-Dagestan oil region was selected from among all oil regions because it shows clearly all three characteristic zones of ground water development. For example, in the Chernye Mountains of the Grozny region, the discharge of fresh springs from the Karagan and Chokrak formations restricts the ground-water recharge zone in the productive formations. In the Sunzha and Terek ranges it is possible to collect water samples from wells where natural movements of the ground water have practically ceased and where, consequently, one should be able to observe concentrations of ground water (high mineralization), on the one hand, and accumulations of organic material that migrated with the ground water, on the other. And, lastly, one may observe thermal springs, for the most part, emerging from these same formations in the anticlinal zones of the ranges; in some places these springs yield water with films of oil, marking zones of ground-water discharge in such localities. It was pointed out in the Introduction that one should not search in a direction vertical to the ground surface, when seeking an understanding of the processes at work in ground water, but down the dip of an aquifer; i.e., he should investigate along the flow of the ground water that is confined to some particular water-bearing unit, formation, series, or any other stratigraphic subdivision.

As far as possible, the samples of water were collected everywhere with this rule in mind. Of course, it would be extremely important to collect water samples also in synclines, such as the Sunzha syncline, but it was not possible to do this since there were no drill holes in the district.

A very important rule of technique would be, also, the collection of water samples along definite vertical planes that cut the aquifers along the direction of their dip. But, this procedure was also practically impossible, since existing springs and drill holes are distributed in a rather haphazard pattern. The collection of water samples was therefore made as far as possible along a zone of some particular breadth extending from the recharge zone to the zone of deepest burial or to the zone of discharge of the water-bearing formation. The rate of movement of

---

*All the preliminary remarks concerning arrangements of investigatory work in this chapter were written by M. E. Al'tovskii.

ground water is an extremely important factor in all processes taking place in the water. It may be sufficient to say that the faster the ground water moves the closer the connection is with the earth's surface, the deeper it is that gases of atmospheric origin will be carried into the shell of sedimentary rocks, the deeper the oxygen boundary will lie, and, lastly, the slower will the substances accompanying the ground water accumulate in any quantity.

We made no investigations on direct determinations of velocity of ground-water flow. But, the higher the yield of a well, obviously the more rapidly the ground water must be moving in that zone. Therefore, we sometimes group our analyses according to wells having yields greater or less than certain amounts.

The velocity of ground-water flow is fundamentally affected by various types of tectonic disturbances (faults), which almost always act as barriers in moderately consolidated strata, stopping the movement of ground water in certain zones.

Therefore, some of the water samples we collected came from wells in which the intake occurred in overthrust blocks or in sole blocks below overthrust plates.

In approaching the study of organic material in ground water, we found ourselves in a very difficult position. The reason for this is that there were no methods for determining organic material in ground waters with high concentrations and specific chemical compositions. Furthermore, the concentrations of organic substances in ground water, in comparison with the other components in the water, are extremely small. Consequently, in order to perform our analytical work, it was necessary to find a technique for extracting and concentrating the organic material from ground water.

This latter task was accomplished by E. L. Bykova as recently as 1956. The extracted and concentrated organic material proved to be very active; it consisted of particles, some positively charged, some negative, and some neutral. Thorough examination of the organic material extracted from the ground water constituted one of the chief tasks in the subsequent analytical work.

Micromethods proved to be the most usable for determining organic carbon and nitrogen. To determine organic carbon we first utilized the micromethod of V. G. Datsko, which he had used for quantitative determinations of carbon in the waters of the Black Sea. This method was tested by A. A. Brodovskii and M. Ya. Dudova on pure organic compounds (saccharose, phenylalanine, and others) and was modified by certain corrections for different substances taking part in the reactions. In this way the apparatus was somewhat improved.

As a result of all the preliminary work on V. G. Datsko's micromethod, it was made clear that the method was usable on ground waters of the most diverse chemical compositions, including brines. The reproducibility of the method amounted to ± 6%. On easily oxidizable substances (carbohydrates), results were obtained that were too low by 17%; on material difficult to oxidize, such as phenylalanine, without the addition of marine salt, the results are too low by as much as 40%, but with the addition of salt, too low by only as much as 20%.

Since most of the analytical work on determining organic carbon in ground water of the Grozny region was done with added marine salt, the data reported below are probably too low by approximately 26%.[*] In 1956 we determined naphthenic acids (works of A. A. Brodovskii and V. N. Belekhova) by the method of A. A. Reznikova and E. P. Mulikovskii. As already reported above, naphthenic acids in the Grozny-Dagestan region are found not only near oil deposits but also in recharge zones. But, naphthenic acids are not now found in the waters from the Carboniferous rocks of the Moscow basin.

The microbiological studies were made by well-known methods. Necessary information on the method of collecting samples is given in Chapter VI. However, it should be noted that all the samples for microbiological investigations were taken from continuously circulating liquids. No samples were collected from stagnant wells or from wells showing poor filtration.

The water samples that were collected, in addition to being studied as indicated above, were subjected to ordinary hydrochemical analysis, to gas analysis, to various tests of oxidizability, to measurements of oxidation-reduction potential, and to examinations for nickel and vanadium content. The data from these investigations will be published after completing all the analytical work on the Grozny-Dagestan oil region.

_____
[*] In 1956, A. A. Brodovskii worked up the micromethod of L. P. Krylova for application to ground water, and obtained almost complete oxidation of organic carbon.

A BRIEF SURVEY OF THE GEOLOGIC STRUCTURE

The gas- and oil-bearing group of rocks includes Cretaceous and Tertiary strata (Fig. 16). The Lower Cretaceous consists chiefly of sandy clay rocks, terminating with a unit of greasy clay rocks of Albian age containing small seams of friable sandstone. The Upper Cretaceous consists predominantly of carbonate sequences — limestones and marls. The total thickness of Cretaceous strata is 2000 m.

The Tertiary section, containing all the series of the Paleogene and Neogene, is made up chiefly of sandy clay rocks. Exceptions to this dominant type are marls and limestones (containing foraminifers) and also layers of Sarmatian and Meotian limestones. The total thickness of Tertiary strata is 5000 m (Sulak River).

The most interesting strata, for our purposes, are the productive Karagan-Chokrak formations of middle Miocene age; they are gas- and oil-bearing strata throughtout the extensive area of eastern Ciscaucasia.

The Chokrak beds consist of alternating dark clays, clay shales, and friable quartz sandstones and sands. Over a large part of the area investigated by us, the lower part of the Chokrak section contains a unit of clay, free of sandstones. The most characteristic feature of the Chokrak strata is its extremely sharp facies variability throughout the area, and also its variation in thickness. Thus, the content of sand increases from west to east, from 0.1 (Malgobek) to 0.4 in the region of the so-called Dagestan wedge; from here it again decreases toward the south, to 0.1-0.2 (Dagestanskie Ogni). The thickness changes in a similar way, increasing from northwest to southeast, from 200 to 1200 m, and then decreasing again to the south of the region of Makhachkala (Khosh-Menzil), to 300 m.

The Karagan beds are similar to the upper sandy clay part of the Chokrak strata, consisting of alternating brown sandy clays, light quartz sands and sandstones, and yellowish gray marls. The basic distinction between the Karagan strata and the Chokrak strata is the greater sand content and the less extensive facies change in the Karagan beds. The thickness changes in the same direction as in the Chokrak strata; i.e., it decreases to the northwest and to the south from 440 to 50-100 m. The Chokrak beds are underlain by a monotonous sequence of stratified, dark, bituminous Maikop clays, containing thin layers of sandstones and marls and having a thickness of up to 1500 m.

The Terek-Dagestan region may be subdivided tectonically into a number of zones.

1. The Cretaceous massif, extending in a belt from the northwest, first to the southeast, and then to the south. From here the rocks disappear under Tertiary strata, to the north and northeast, descending to great depth.

2. North of the belt of Cretaceous rocks occurs the Chernye Mountains homocline, composed of Tertiary strata, which are also buried on the north.

3. The first anticlinal zone, extending from the Stavropol Plateau to the Azerbaidzhan border. This zone consists of a series of separate uplifts (from northwest to southeast): the Sunzha anticlinorium, the Benoi doubly plunging anticline, the Cretaceous dome of Sala-Tau and El'dema, the Gasha and Bolkhas-Khunuk doubly plunging anticlines, and, in the southernmost part, the Adzhinaur uplift.

4. The second anticlinal zone is parallel to the first. It, like the first, consists of a number of uplifts, following one another in sequence: the Terek anticlinorium, the Gudermes doubly plunging anticline, the Cretaceous domes of Khadum and Kukurt-tau, and the doubly plunging anticlines of Auch-Su, Izberbash, Kayakent, Berekei, Dagestanskie Ogni, Rukel', and Khosh-Menzil.

5. Both anticlinal zones are divided up by trough-like synclinal basins.

6. The broad Ciscaucasian depression is found in the Grozny region, north of the Terek anticlinorium, but in the Dagestan region, east of the second anticlinal zone, the Tertiary rocks descent toward the Caspian Sea.

All the principal oil and gas deposits are confined to the above-indicated two anticlinal zones. The uplifts in the anticlinal zones are chiefly short anticlinal structures, box-like, with broad and gently inclined crests and steep limbs. They are all complicated by faults having great vertical displacement (up to 1200 m) and small horizontal displacement. The faults most commonly cut the folds parallel to the fold axes. The folds in the Grozny region have been extremely complicated by faults.

Commercial deposits of oil in the Grozny region are confined to the Chokrak and Karagan strata; in the northern part of Dagestan, oil deposits occur only in the Chokrak strata and, in part, in Maikop beds, whereas in southern Dagestan oil and gas are confined to the Khadum horizon and to Lower Cretaceous strata.* In addition, commercial deposits of gas in the Grozny region are found in the Foraminiferal beds, the Khadum horizon, and upper Sarmatian strata, and signs of oil are observed throughout the section of Tertiary rocks.

Thus, it may be said that gas and oil are present to some degree in all the deposits, beginning with the Lower Cretaceous and extending through to the upper Sarmatian.

GENERAL HYDROGEOLOGIC CONDITIONS

In the Grozny and Dagestan oil region there are rather well-defined conditions of recharge, movement, and discharge of ground water. In this connection, our investigations were made on the basis of already existing schemes of distribution of ground water in strata according to chemical composition, temperature, direction of flow, and other peculiarities. But, at the same time, we obtained some new information, which we also used in the present work.

The recharge zone of the water-bearing horizons for the productive beds of the Karagan and Chokrak strata in the Grozny region are in the Chernye Mountains (Chernye Mountains homocline); in Dagestan, they are in the foothills. The water moves from the recharge zone to the zone of oil and gas deposits, first passing through deep synclinal depressions. Thus, in the Grozny region, the Chokrak strata descend to a depth of 4000 m in the Sunzha syncline, and in northern Dagestan the same strata descend to 300 m; in southern Dagestan Cretaceous strata also descend to depths of 4000 m.

A zone of partial discharge for the water-bearing horizons is also found in regions of anticlinal uplifts, where the productive beds outcrop.

Below we will consider briefly only those horizons that were studied by us for their content of organic material.

Waters of the Karagan Strata

Fresh cold springs are found in the recharge zone for these strata. The temperature of the water is 12-19°C. Mineralization ranges from 0.4 to 1.3 g/liter. Chemically

---

*Recently commercial deposits of oil have been discovered in Upper Cretaceous deposits.

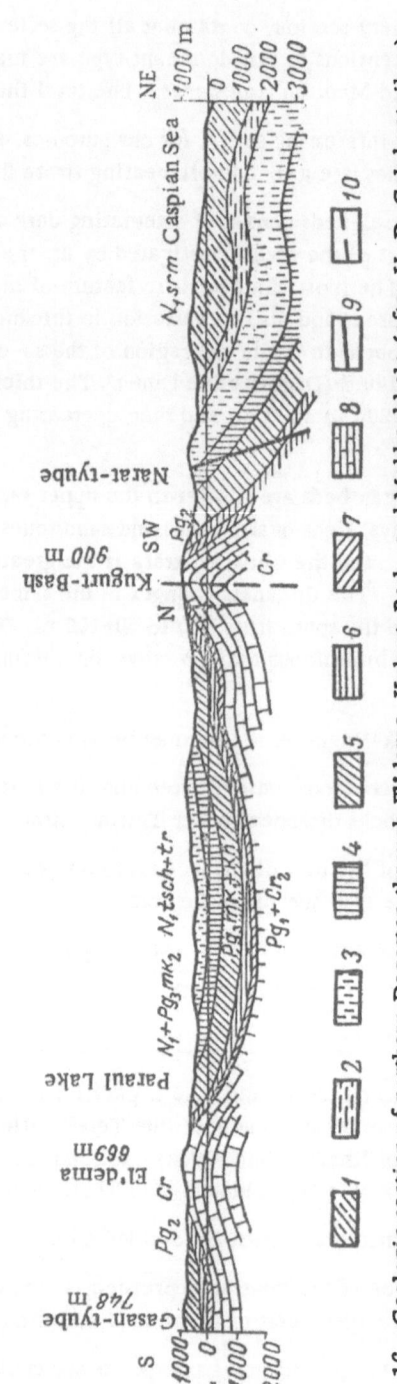

Fig. 16. Geologic section of northern Dagestan through El'dema, Kukurt-Bash, and Makhachkala (after V. D. Golubyatnikov). 1) Sarmatian strata, 2) Konka and Karagan horizons, 3) Chokrak and Tarkhan horizons, 4) Upper Maikop series, 5) Lower Maikop series and the Khadum horizon, 6) Foraminiferal beds, 7) Pestrotsvetnaya (mottled) series, 8) Cretaceous rocks, 9) fault, 10) axis of bending in the line of section.

the waters are chiefly three-, four-, and multicomponent* types, the HCO_3^- and Ca^{++} ions predominating. The springs do not yield more than 0.5 liter/sec.

In the Grozny region, waters in the Karaganda strata near oil deposits are found at depths ranging from 350 to 1000 m, locally producing artesian flow at the surface; at the Makhachkala field in Dagestan, these waters flow freely from wells with depths ranging from 600 to 1600 m. The water temperatures are rather high, from 30 to 85°C. The chemical composition of the waters in the Karagan beds of the Grozny region is somewhat different from that of waters in the same beds in the Dagestan region. In the first region, the waters show mineralization on the order of 1-2 g/liter with two- and three-component composition: sodium bicarbonate-chloride and, rarely, sodium chloride-bicarbonate. In Dagestan the waters are more highly mineralized (from 3 to 6 g/liter), sulfate forming the dominant anion (sodium sulfate-bicarbonate type).

The yield of the wells is quite variable, from tens to hundreds of cubic meters per day. The yield of well No. 160 in Makhachkala is phenomenal, amounting to 1700 cubic meters of mineral water per day.

Waters from springs in the discharge zones differ markedly in their chemical composition in the Grozny and Dagestan regions. In the Grozny region, the waters are but weakly mineralized, 1-2 g/liter, of two-, three-, and four-component types, chiefly sodium bicarbonate-sulfate. In Dagestan the waters are rather highly mineralized, 30-40 g/liter, of two-component sodium chloride type. All the springs contain hydrogen sulfide. Water temperatures range from 12 to 45°C, and the yield of the springs ranges from hundreds of liters down to tenths of a liter per second.

Waters of the Chokrak Strata

In recharge zones occur chiefly fresh, cold springs with mineralization not exceeding 1 g/liter, chemically resembling the Karagan waters. But, in the Dagestan region highly mineralized waters are found in addition to the fresh waters, having a mineralization of up to 3 g/liter, in places of sodium chloride type, but more commonly of sodium sulfate-bicarbonate type. Even more highly mineralized waters are found in the Datykh district (Chernye Mountains), where the mineralization reaches 150 g/liter. All these peculiarities in chemical composition of waters in the recharge zone are undoubtedly related directly to the lithology of the water-bearing horizons. Thus, in the Datykh region the Chokrak beds are characterized by the presence of saline-gypsiferous layers.

Waters with quite variable mineralization (from 0.6 to 28 g/liter) are encountered in the vicinity of oil deposits. A brief sketch of the basic pattern in chemical composition is as follows: a) at deposits in folds modified by faults (including most of the deposits in the Grozny region), the water in any particular bed is generally more highly mineralized in the part of the structure beneath the thrust plane; b) waters near the oil-water contact of oil deposits are generally more highly mineralized than waters found some distance from the contact; c) waters in water-bearing (and oil-bearing) horizons of low permeability (either because of lithology or facies changes) have higher mineralization than waters in beds of higher permeability.

Waters in the Chokrak beds are found at oil deposits at greater depth than the Karagan beds (on the order of 1-2 km, except for the Achi-Su deposit), and they therefore have higher temperatures (from 50 to 100°C). According to computation, the maximum temperature reaches 122°C (in bed XXII at the Oktyabr'skoe deposit). There is a definite relationship between water temperature and mineralization, waters in zones of highest temperatures having the highest mineralization.

In most of the wells the liquid is pumped out, but in some regions there are still flowing wells (mostly giving pure water, with no petroleum). The yield in these wells ranges widely, from tens (and even less) to several hundreds of cubic meters per day. In water wells the yield is almost always many times greater, commonly amounting to thousands of cubic meters per day. For example, the well at Gudermes (Kokhanovka No. 1) gives approximately 4000 cubic meters per day.

---

*Two-component systems include waters in which two cations and two anions are present in quantities less than $20 \pm 3\%$ mg-equiv. each.

Three-component types are waters in which the content of three components out of six are chiefly less than $20 \pm 3\%$ mg-equiv. each. Four-component types are those in which two components out of six are present in quantities less than $20 \pm 3\%$ mg-equiv. each.

The remaining types of waters are called multicomponent or mixed. The total of cations and anions is taken to be 100% for each separately.

A rather large number of mineral springs occur in the discharge zone, most of them used for medicinal purposes. The best known of these springs are the Sernovodsk, Miatli, Bragun, Isti-Su, Kayakent, and others. The springs of "Goryachii Klyuch" (Isti-Su) are of most interest; at these springs films and clots of oil emerge with the water, although no oil deposit is found near them.

Most of the water in these springs, is hot, on the order of 40-90°C, but there are some springs with water temperatures of 20-30°C. Chemically the waters are also extremely variable; the mineralization ranges from 0.8 to 28 g/liter. Relatively weakly mineralized waters are chiefly of the sodium bicarbonate-sulfate-chloride type, but highly mineralized waters are sodium chloride-bicarbonate and sodium sulfate-bicarbonate types. Two-component waters are much more rarely encountered: sodium chloride and sodium sulfate types. The discharge at these springs is small for the most part, generally ranging from 0.005 to 1.5 liter/sec, most commonly tenths of a liter per second. Some springs, formerly flowing (such as the Goryachevodskaya and others) have completely dried up in recent years, and others have diminished in flow, a fact explained by extensive extraction of liquid at the oil fields.

Waters of the Maikop Beds

Despite the fact that the Maikop beds are chiefly clay rocks, they contain sandy layers and fractures along which waters circulate. In the Chernye Mountains and in the Dagestan foothills the Maikop strata supply springs with fresh and weakly mineralized waters (up to 3 g/liter), yields of tenths of a liter per second, and water temperatures of 15-18°C. In addition, flowing artesian wells, from exploratory drill holes, are found in the Chernye Mountains, with yields up to 700 cubic meters per day, water temperatures up to 58°C, and mineralization in excess of 30 g/liter (of the sodium chloride type). The artesian waters are accompanied by abundant elimination of methane. In the Terek Range, when sampling the Lower Maikop beds at a depth of 3305-3308 m, a flowing well was obtained, with a yield of up to 5 cubic meters per day of water of the sodium chloride type (mineralization of 17 g/liter) and having a maximum pressure at the collar of 145 atm. In northern Dagestan the Maikop beds also contain saline waters, with a mineralization of 30 g/liter (of the sodium chloride type). In southern Dagestan the Maikop beds are clays with layers of siltstones, from which a considerable flow of water is obtained in several wells. The water here circulates chiefly along joints. Mineralization is 50-96 g/liter. The water is of the sodium chloride type.

Waters of the Khadum, Foraminiferal, and Upper Cretaceous Strata

These beds are chiefly carbonate rocks, with no impermeable layers separating them; they therefore probably constitute a single water-bearing horizon. These rocks crop out at the surface in the Chernye Mountains and in the Dagestan foothills, where they yield fresh and slightly mineralized waters with temperatures of 5-12°C. Some springs with fresh cold water are characterized by the presence of H_2S.

The highest-yield springs with the lowest mineralization are found in the Upper Cretaceous limestones. Thus, a spring near the village of Bol'shoe Kazanishche (northern Dagestan) has a yield of several hundred liters per second. Its water is of the calcium bicarbonate type and has a mineralization of 0.2 g/liter. The Foraminiferal beds in the Chernye Mountains yield, from wells, saline waters of the sodium chloride type, with mineralization up to 45 g/liter and temperatures of 50°C. At the Talgi resort in Dagestan, the Foraminiferal beds supply water of average mineralization (from 5 to 7 g/liter) of the sodium-calcium chloride type with a temperature of 31-38°C. The H_2S content is on the order of 400-500 g/liter. In the vicinity of deposits, waters of the Khadum, Foraminiferal, and Upper Cretaceous strata have been studied only at the gas deposits of southern Dagestan. These waters are saturated with hydrocarbon gases, CO_2, and N_2; the mineralization is 70-80 g/liter and the temperature 30-40°C. The yield in wells here is measured in tens of cubic meters per day. Formerly some of these wells flowed freely, with yields up to several thousands of cubic meters per day.

Waters of Lower Cretaceous Strata

These waters were studied in southern Dagestan. Here the Lower Cretaceous consists of Aptian-Albian clays, argillaceous siltstones, and siltstones. Chemically the waters are of the sodium chloride type, with mineralization of 67-77 g/liter.

Determination of the Oxidation-Reduction Potential

We determined the physicochemical conditions in the aquifers by measuring the oxidation-reduction potential (data of Z. I. Kuznetsova). These measurements were made by an electrometer and the values were reduced to a normal hydrogen electrode. The results show that an oxidation environment exists in waters at the recharge zone. Here the value of Eh ranges from $+197$ to $+525$ mv (the average value is on the order of $+200-300$ mv). In the vicinity of oil deposits conditions change, the value of Eh ranging from $+42$ to -17.7, negative values predominating. In the vicinity of gas deposits there is also a change of conditions (Eh from -29 to $+132$ mv), but here positive values predominate. The negative values of Eh range from -25 to -29 mv. Only negative values of Eh were obtained in discharge zones (from -16 to -204 mv), a fact indicating that reducing conditions exist in the waters at the discharge zone.

A Summary of the Geologic History of the Region

I. M. Gubkin has distinguished the following stages in the historical geology of the region. The Cretaceous Sea transgressed, and, at the end of the transgression, a thick carbonate sequence was deposited; in Maestrichtian time uplift began and, because of this, the sea grew shallow; at the beginning of the Tertiary, as a result of later Laramide movements, the region was depressed again; during the Eocene and Oligocene, large-scale mountain-making movements occurred, leading to the formation of the principal anticlinal uplifts in the region. After these movements there followed a long period of sediment accumulation in the near-shore zone along the north slope of the Caucasus. After this, accompanied by fluctuations of the shore line, there was deposited a whole sequence of sediments, from the lower part of the Maikop series to the upper Sarmatian series inclusive. After the late Sarmatian stage and before the Akchagyl transgression all the tectonic elements now found in the region were finally formed. During Akchagyl time there was a transgresssion of the sea, and this interval ended with a new rise of dry land.

Let us pause for somewhat more detailed examination of each stage. Before the Pyrenees phase of folding there was a time of quiet accumulation of the Cretaceous deposits. Intense tectonic movements gave rise to the extensive uplift of the Caucasus and, as a result, carbonate sediments give way to clastics. At the end of the Miocene, tectonic movements finally ceased. Whereas in Maikop time the Caucasian Range formed a chain of small islands, at the beginning of Chokrak time it became a single large island. The time of accumulation of the Chokrak-Karagan sequence is subdivided into two intervals: the first,when a clay sequence was deposited, and the second,when a clay-sand sequence accumulated. During the second interval there were frequent fluctuations of the shore line, which led to alternating accumulations of clay and sandy units. Already at this time, when the sea periodically withdrew, sandy beds that were exposed at the surface were receiving meteoric water, which began to move down the dip of the beds. Beginning with Konka time, and continuing until Meotian time, the sea covered the region. A new uplift of the Caucasian land mass occurred at the beginning of Meotian time. Folding occurred at that time, leading to the formation of the Peredovye Ranges and the Tertiary foothills of Dagestan. At the beginning of the Akchagyl transgression these uplifts had undergone considerable erosion, as a result of which the Karagan-Chokrak and, probably, older strata appeared locally at the surface, where they have served as windows for discharge of groundwater. Thus, as early as at the beginning of Akchagyl time, there might have existed a hydrodynamic connection between zones of recharge and zones of discharge. This connection was probably not destroyed during the Akchagyl transgression, since the regions near the anticlinal crests were not covered by the sea but remained as small islands. At the end of Apsheron time, uplift occurred over the entire region, accompanied by intense folding movements, complicating the box-like anticlines that had formed earlier (in the pre-Akchagyl phase). All the principal faults of the region formed at this time. After the Apsheron uplift the sea never again returned to any of the described region.

It should be added to what has been said above that, as an example, if we consider the porosity of oil-bearing sandstones to be a minimum (10%), the length of time for complete water exchange in these beds would amount to about 4000 years. Therefore, for a period of but half a million years, i.e., approximately for a time equivalent to the Quaternary period, the store of water in these beds must have been renewed no less than 125 times. From this it is difficult to conceive how sedimentational water of marine origin can be preserved in sandy beds, no matter how slight their permeability.

The Content of Organic Material in Ground Water

The field work on collecting water samples for determining organic carbon was done by us in 1954-1955.[*]

During the first year (June-August) the Grozny region was studied; the Dagestan region was studied in the second year (September-October). We collected 152 samples of water from springs and wells for both field periods. The region of investigation included an area from the Chernye Mountains and the Dagestan foothills on the south and southwest to the Terek Range and the Caspian Sea on the north and east. We studied, in various detail it is true, waters from all the large oil and gas fields: Malgobek-Voznesenskoe, Gora-Gorskaya, Tashkala, Starogrozny, Oktyabr'skoe, Novogrozny, Makhachkala, Achi-Su, Izberbash, Duzlak, Dagestanskie Ogni, and Khosh-Menzil.

From the zone of recharge 29 samples were collected. All these were taken from springs, mostly by tapping. From areas about oil or gas deposits, 84 samples were collected. Some of these samples were taken from pumped wells, giving water and oil; some were from flowing wells, giving pure water for the most part. The collection of water from oil wells was generally made from test cocks on discharge pipes connecting the collars of wells with settling tanks. Samples were collected at wells that had not stood idle prior to collecting the water, but had been active without pause for long periods of time. We collected 39 water samples from flowing springs at the zone of discharge. Some of the springs had cemented-masonry collecting receptacles, but most were tapped in no way.

We collected 4 samples from the recharge zone of the Karagan strata. The results of the analyses are shown in Table 17.[**]

We have included data from 17 analyses of waters from regions of oil deposits. Most of these are for waters collected at the Obtyabr'skoe field.

In Table 18 we show the results of some analyses of water to indicate the minimum and maximum content of organic carbon and the average for 8 analyses for two water groups. The first group includes water collected either from high-yield wells giving emulsions or from flowing wells giving pure water; the second group includes water taken from low-yield wells giving emulsions.

In regard to the relationship between content of organic carbon and the structural features at the sample locality, the depth to the aquifer, mineralization, chemical composition, and temperature of water, and the content of oil, it is not yet possible to draw any sort of convincing conclusions, because of the still insufficient number of analytical data. However, in analyzing what has been said above, the following remarks may be made:

a) no systematic change of content of organic carbon with depth to aquifer (along the vertical) was observed at the Oktyabr'skoe or Starogrozny fields;

b) neither was any relationship between content or organic carbon and mineralization or chemical composition of the water detected. For example, oil-free water at the Starogrozny field differs sharply in degree of mineralization from such water at the Makhachkala field (the mineralization at Makhachkala is 4-5 times that at the Starogrozny field), but the content of organic carbon is practically the same in both (on the order of 1-3 mg/liter).

Six analyses were made of waters in the Karagan beds at the recharge zone. The results are given in Table 19.

As these data show (in Table 19), all the waters contain small quantities of organic carbon (<10 mg/liter). It should be remarked here that the smallest content of carbon is found in water with the lowest mineralization and the highest temperature (sample 34/54).

In 1956, we determined, besides organic carbon, naphthenic acids as well. In waters from the recharge zone of the Karagan beds, naphthenic acids were detected in only one of seven samples; the content in that one was 0.6 mg/liter. This sample was taken in a spring emerging on the outskirts of a village, the water collecting in rock. The water was probably contaminated, since the content of NO_3^- is very high (44 mg/liter).

[*]The analyses were made by M. Ya. Dudova and A. A. Brodovskii.

[**]The data for 1954 and 1955 are shown in Tables 17, 18, 19, 20, 21, 22, 23, 24, 25, and 26.

Results of Analyses on Waters from the Karagan Strata

Sample No.	Sample locality	Discharge of spring, liter/sec	Water temp., °C	Chemical character	Content of C_{org}, mg/liter	Remarks
72/54	Benoi	—	19.0	$M_{0,4} \dfrac{HCO_{84}^3 SO_{13}^4 Cl_3}{Ca_{74} Mg_{24} Na_2}$	4.4	Well 0.5 m deep
70/54	Benoi	0.06	12.5	$M_{1,3} \dfrac{SO_{56}^4 HCO_{42}^3 Cl_2}{Ca_{55} Mg_{42} Na_3}$	5.0	Tapped spring
13/55	Agach-aul	0.07	16.0	$M_{1,2} \dfrac{HCO_{39}^3 SO_{33}^4 Cl_{28}}{Ca_{61} Mg_{35} Na_4}$	0.9	Tapped spring
63/55	Darvak	0.15	12.0	$M_{0,7} \dfrac{HCO_{62}^3 NO_{16}^3 Cl_{12} SO_{10}^4}{Ca_{44} Na_{33} Mg_{25}}$	1.2	Tapped spring $NO_3' = 120$ mg/liter $NO_2' = 0.12$ mg/liter
				Average of 4 samples	2.9	

Results of Analyses on Waters with Minimum and Maximum Content of Organic Carbon

Sample No.	Sample locality	Designation of bed	Position of well in the structure	Perforated interval	Yield of liquid, tons/day	Content of oil, %	Water temp., °C	Chemical character	Content of C_{org}, mg/liter
12/55	Makhachkala field, well 160	—	Northeastern part of the buried northern limb of a doubly plunging anticline	1560	1700	None	63.0	$M_{5,0} \dfrac{SO_{42}^4 HCO_{31}^3 Cl_{28}}{Na_{98} Ca, Mg}$	0.9 (min)
10/54	Oktyabr'skoe field, 2nd subsidiary firm	I	Crest of fold	350—358	474.8	0.1	60.0	$M_{2,0} \dfrac{HCO_{56}^3 Cl_{41} SO_3^4}{Na_{96} Ca_2 Mg_2}$	10.2 (max)
								Average of 8 analyses	4.5
14/54	Oktyabr'skoe field, 1st subsidiary firm	XIII	Northern zone near anticlinal crest	767.7 — 770.8	295.8	0.2	85.0	$M_{0,9} \dfrac{HCO_{56}^3 SO_{28}^4 Cl_{16}}{Na_{92} Ca_6 Mg_2}$	11.4 (min)
8/54	Okryabr'skoe field, 2nd subsidiary firm	XII	Southern zone near anticlinal crest	547—557	52.3	2.5	45.0	$M_{2,3} \dfrac{HCO_{76}^3 Cl_{20} SO_4^4}{Na_{95} Ca_3 Mg_2}$	163.0 (max)
								Average of 8 analyses	51.4

TABLE 19

Analyses of Waters in the Discharge Zone of Karagan Strata

Sample No.	Sample locality	Discharge of spring, liter/sec	Water temp., °C	Chemical character	Content of C_{org}, mg/liter	Remarks
64/54	Sernovodsk, iron spring	0.04	20.5	$M_{1.8} \dfrac{HCO_{73}^3 SO_{24}^4 Cl_3}{Na_{94} Mg_4 Ca_2}$	6.4	Tapped spring
58/54	Sleptsovka spring	0.5	21.5	$M_{1.3} \dfrac{SO_{49}^4 HCO_{42}^3 Cl_9}{Na_{92} Ca_4 Mg_4}$	3.5	Tapped spring
57/54	Upper Achaluki	—	12.5	$M_{2.1} \dfrac{Cl_{42} HCO_{31}^3 SO_{27}^4}{Na_{75} Mg_{13} Ca_{12}}$	5.2	Well 80 m deep
56/54	Middle Achaluki	0.2	25.0	$M_{1.3} \dfrac{HCO_{81}^3 SO_{11}^4 Cl_8}{Na_{96} Mg_3 Ca_1}$	3.7	Tapped spring
33/54	Bragun spring No. 5	0.25	35.0	$M_{2.2} \dfrac{HCO_{75}^3 SO_{16}^4 Cl_9}{Na_{98} Ca_{1.5} Mg_{c.5}}$	4.4	Odor of oil
34/54	Bragun spring No. 6	0.25	44.5	$M_{1.2} \dfrac{HCO_{40}^3 SO_{39}^4 Cl_{21}}{Na_{97} Ca_2 Mg_1}$	1.3	—
				Average of 6 samples	4.1	

TABLE 20

Analyses of Waters from Chokrak Strata

Sample No.	Sample locality	Discharge of spring, liters/sec	Water temp., °C	Chemical character	Content of C_{org}, mg/liter	Remarks
29/55	Buinaksk	0.08	14.0	$M_{0.6} \dfrac{SO_{46}^4 HCO_{46}^3 Cl_6 NO_2^3}{Ca_{58} Mg_{24} Na_{18}}$	0.9 (min)	Tapped spring NO_3' — 12 mg/liter NO_2' — none
69/54	1.5 km southeast of Betli	0.1	13.5	$M_{0.8} \dfrac{HCO_{63}^3 SO_{36}^4 Cl_1}{Mg_{53} Ca_{47}}$	11.1 (max)	Tapped spring NO_3' — none NO_2' — none
				Average of 13 analyses	2.8	

TABLE 21

Comparison between Content of Organic Carbon and Chemical Composition of the Water

Sample No.	Sample locality	Discharge of spring, liters/sec	Water temp., °C	Chemical character	Content of C_{org}, mg/liter	Remarks
13/54	Kydyrovo	—	15.0	$M_{0.08} \dfrac{SO_{44}^4 HCO_{31}^3 NO_{14}^3 Cl_{11}}{Ca_{48} Mg_{31} Na_{21}}$	1.5	Nonflowing water NO_3' — 9 mg/liter NO_2' — not determined
45/55	0.5 km west of Ersi	—	15.0	$M_{2.3} \dfrac{Cl_{51} SO_{31}^4 HCO_{18}^3}{Na_{82} Ca_{13} Mg_5}$	1.6	Nonflowing water NO_3' — none NO_2' — none

In the vicinity of an oil or gas well, all the water in the Karagan beds contains naphthenic acids. The content varies: in waters from flowing wells (not yielding oil) the content is 0.1-0.2 mg/liter; in water from wells yielding emulsion, the value is 1-3 mg/liter. Waters from springs at the discharge zone of the Karagan beds also contain naphthenic acids. The maximum quantity is 2.0 mg/liter, the minimum 0.3 mg/liter. In but one sample of five did we fail to detect naphthenic acids.

The largest amount of water for analysis was taken from the Chokrak strata. We collected 13 samples from springs at the recharge zone. Many springs are found in inhabited localities, and some of these are contaminated to some degree (judging from the content of nitrogenous compounds). But, regardless of this, all the waters contain approximate identical quantities of dissolved organic material, as seen in Table 20, where we have shown the minimum, maximum, and average content of organic carbon.

The existing data permit us to conclude that there is no relationship between the content of organic material and the chemical composition of the water, as is confirmed, to some degree, by the data in Table 21.

As the data in Table 21 show, the water does not flow in either spring (the samples were collected in pails), the temperature is the same in both, but the mineralization and chemical composition differ markedly in the two. However, the content of organic carbon is almost the same in both.

From the vicinity of oil deposits, 41 samples were collected of waters in the Chokrak strata. The samples were taken, as far as possible, to represent waters of the principal productive horizons. At the same time, samples were taken from the wells yielding the highest percent of water.

Table 22 presents the results of water analyses from the Chokrak beds (giving minimum, maximum, and average content of organic carbon for the above-distinguished two water groups).

In making a greater number of analyses of waters from the Chokrak beds than from the Karagan beds, we may here compare the data on content of organic carbon with some of the more important (in our view) geologic and hydrogeologic factors. Our preliminary conclusions on this matter are outlined briefly in the following.

1. The absence of any relationship between content of organic carbon and chemical composition of the water, mentioned earlier, is confirmed. Neither have we observed any systematic change in the content of organic carbon with vertical position in the section.

2. We did not succeed in finding any distinct pattern in the distribution of organic carbon in relation to structural features in the aquifers. However, beds beneath overthrust plates have more organic material than beds in the thrust block.

3. We detected no direct relationship between content of organic carbon in water and the percent of oil in the water, from which one might suppose that ground water here is enriched in organic carbon by the destruction of oil deposits (see Table 23).

The data in Table 23 shows that no direct connection is actually observed between content of oil in extracted emulsions and the content of organic carbon in the water. Thus, in the first example, in sample 5/54, oil is somewhat less abundant than in sample 17/54, though the organic carbon in the first sample is three times that in the second. In the second example, in sample 79/55, oil is twice as abundant as in sample 74/55; nevertheless, organic carbon is more abundant in the sample with the lower content of oil. There are also several less graphic examples that support the position we have just stated. But there are other examples, though much rarer, in which the higher oil content in water corresponds to a higher content of organic carbon.

4. Waters in deposits of the Sunzha Range have a higher content of organic carbon than waters at deposits in the Terek and Gudermes ranges, and higher yet in comparison with waters at deposits in Dagestan.

5. The minimum content of organic carbon was found in waters from wells situated beyond the edge of oil zones. Such wells, as a rule, are freely flowing, with pure oil-free water.

6. A low content of organic carbon was also noted in waters from some wells, even though within the oil-bearing zone and yielding some quantity of oil, which were mostly flowing wells or, if not flowing, were high-yield wells. Except for this, there are only individual examples of water from Dagestan oil fields where a low content of organic carbon comes from wells pumping very small quantities of emulsion (a few cubic meters per day), although the percent content of oil is high (up to 25%).

TABLE 22
Analyses of Waters from Chokrak Beds in the Vicinity of Oil Deposits

Sample No.	Sample locality	Designation of bed	Position of well in the structure	Perforated interval, m	Yield of liquid, tons/day	Content of oil, %	Water temp., °C	Chemical character	Content of C_{org}, mg/liter
3/55	Makhachkala field, flowing well 27	Series "B"	Axial part of the periclinal termination of a fold	1505—1509	650	None	56.0	$M_{1.8}\dfrac{HCO_{48}^3 SO_{40}^4 Cl_{12}}{Na_{98}Ca_1Mg_1}$	0.5 (min)
4/54	Okryabr'skoe field, 2nd subsidiary firm, flowing well 45/13	XXII	Crest of the fold	838—860	1263	0.2	76.0	$M_{1.8}\dfrac{HCO_{49}^3 Cl_{35} SO_{16}^4}{Na_{98}Ca_1Mg_1}$	9.8 (max)
								Average of 20 analyses	3.8
10/55	Ternair, well 10	Series "B"	Western part of a homocline	1646—48	49	4	48.0	$M_{4.5}\dfrac{HCO_{49}^3 Cl_{48} CO_2^3}{Na_{98}Ca_{0.6}Mg_{0.5}}$	10.5 (min)
5/54	Oktyabr'skoe field, 2nd subsidiary firm, well 29/13	XVI	Crest of the fold	639	176.3	1.3	60.0	$M_{1.8}\dfrac{Cl_{41}HCO_{33}^3 SO_{26}^4}{Na_{98}Ca_4Mg}$	>81 (max)
								Average of 21 analyses	27.7

TABLE 23
Comparison of Content of Organic Carbon and the Percent Content of Oil Water

Sample No.	Sample locality	Designation of bed	Position of well in the structure	Perforated interval, m	Yield of liquid, tons/day	Content of oil, %	Water temp., °C	Chemical character	Content of C_{org}, mg/liter
5/54	Oktyabr'skoe field, 2nd subsidiary firm, well 29/13	XVI	Crest of the fold	639	176.3	1.3	60.0	$M_{1.8}\dfrac{Cl_{41}HCO_{33}^3 SO_{26}^4}{Na_{98}Ca_4Mg_1}$	>81
17/54	Oktyabr'skoe field, 1st subsidiary firm, well 92/20	XVI	Crest of the fold	777—78	322.5	1.7	77.0	$M_{1.1}\dfrac{HCO_{44}^3 SO_{33}^4 Cl_{23}}{Na_{94}Ca_3Mg_3}$	27.8
74/55	Izberbash, well 149	"D" I—II—III	Northeastern part of domal fold	1530—40 1612—35 1682	16.8	10.0	29	$M_{27.6}\dfrac{Cl_{94}HCO_6^3}{Na_{99}Mg_1}$	59.0
79/55	Izberbash, well 186	"D" II	Central part of the dome	1523—40	9.5	20.0	21	$M_{27.1}\dfrac{Cl_{92}HCO_8^3}{Na_{98}Ca_1Mg_1}$	43.0

TABLE 24

Analyses of Waters from Chokrak Beds in the Discharge Zone

Sample No.	Sample locality	Position of well in the structure	Perforated interval, m	Chemical character	Content of C_{org}, mg/liter	Remarks
21/54	Goryachii Klyuch, group III (eastern oil group)	0.8	63.0	$M_{1.6} \dfrac{HCO_{77}^3 Cl_{20} SO_3^4}{Na_{97} Mg_2 Ca_1}$	2.0 (min)	Films and clots of oil emerge with the water
52/55	Dagestan, 3.5 km east of Utemysh	—	21.0	$M_{15.7} \dfrac{SO_{62}^4 Cl_{32} HCO_4^3}{Na_{97} Ca_2 Mg_1}$	17.4 (max)	Many small seeps of water, forming a swampy area
				Average of 19 analyses	4.4	

TABLE 25

Analyses of Waters from Maikop Beds

Sample No.	Sample locality	Position of well in the structure	Perforated interval, m	Yield of wells, tons/day, of springs, liters/sec	Content of oil, %	Water temp., °C	Chemical character	Content of C_{org}, mg/liter	Remarks
54/44	Srednii Datykh, upper reaches of a creek flowing into the Fortanga River	—	—	0.1	None	19.0	$M_{11.8} \dfrac{Cl_{71} SO_{26}^4 HCO_3^3}{Na_{62} Ca_{24} Mg_{14}}$	9.0	Samples taken after a rain
57/55	Dagestan, Chumli, a well	—	—	—	None	14.0	$M_{2.9} \dfrac{SO_{43}^4 HCO_{29}^3 Cl_{28}}{Na_{40} Mg_{34} Ca_{26}}$	6.3	Well 40 m deep. Water taken 4 m from the earth's surf.
52/54	Exploratory district of Gandal-Bas, well 4	Dome of the Datykh anticline	406.3	8.5	Films of oil, emanations of gas	46.5	$M_{31.8} \dfrac{Cl_{98} HCO_2^3}{Na_{91} Ca_5 Mg_1}$	5.2	Flowing
27/55	Northern Dagestan, Leninakent, well 4	Homocline	928—987	4.2	40	18.0	$M_{30.0} \dfrac{Cl_{84} HCO_{16}^3}{Na_{95} Mg_4 Ca_1}$	52.2	Pumped well

TABLE 26

Analyses of Waters from the Khadum Beds at Gas Deposits

Sample No.	Sample locality	Position of well in the structure	Depth of well, m	Yield of water, m³/day	Yield of gas, m³/day	Water temp., °C	Chemical character	Content of C_org, mg/liter
47/55	Dagestanskie Ogni, well 17	Central part of the eastern limb	257	28	3100	39.0	$M_{68.2}\dfrac{Cl_{98}HCO_2^3}{Na_{92}Ca_5Mg_3}$	3.3
46/55	Dagestanskie Ogni, well 37	Southeastern part of the eastern limb	271.7	22	4000	39.0	$M_{69.6}\dfrac{Cl_{98}HCO_2^3}{Na_{92}Ca_5Mg_3}$	2.3
50/55	Dagestanskie Ogni, well 19	Northern periclinal termination	No data	15—17	No data	41.0	$M_{68.9}\dfrac{Cl_{98}HCO_2^3}{Na_{92}Ca_4Mg_3}$	4.2
37/55	Khosh-Menzil, well 11	Central part of the eastern limb	354	27	1000	23.5	$M_{82.7}\dfrac{Cl_{99}SO_1^4}{Na_{92}Mg_4Ca_4}$	8.1
							Average of 4 samples	4.5

From the discharge zone for waters of the Chokrak beds, we collected samples from all the well-known springs in the Grozny region and in Dagestan. In all, 19 samples of water were studied. All the water, except for that in two samples, contains a normal (small) quantity of organic carbon. A high content of carbon in these two samples might be explained by the fact that the waters at these localities emerge in swampy districts, where they ooze out with very low discharge. In addition, these waters are distinguished from the waters of other springs by high mineralization (up to 28 g/liter). The results of the analyses of waters from the discharge zone are shown in Table 24.

We should turn our attention to the fact that a content of organic carbon in greater than average amounts is found in waters of higher mineralization. For example, waters with mineralizations of 2.8, 6.7, 6.4, 27.7, and 15.7 g/liter contain, respectively, 5.6, 5.2, 6.7, 13.4, and 17.4 mg/liter organic carbon.

Naphthenic acids are found in waters of all zones in the Chokrak strata. In the recharge zone, naphthenic acids were detected in 5 of 12 samples, in amounts ranging from 0.1 to 1.0 mg/liter. In waters from the vicinity of oil deposits, collected from wells yielding petroleum emulsion, the content of naphthenic acids ranged from 0.8 to 4.0 mg/liter, the average being 1.4 mg/liter. Waters from wells yielding pure oil-free water do not always contain naphthenic acids. Waters from almost all the springs in the discharge zone contain naphthenic acids, in amounts ranging from 0.4 to 2.0 mg/liter. The average value (taken from 15 analyses) is 0.8 mg/liter. Thus, we may see that naphthenic acids are present in waters of all zones, but in different percentages of samples and in various quantities within each zone.

Waters in the Maikop beds were studied by us very inadequately. In all, there are 4 analyses, the results of which are shown in Table 25.

From the data in Table 25 it may be seen that the largest quantity of organic carbon is found in waters from oil wells at Leninakent. But, at the same time, another (sample 52/54) contains very little organic carbon despite the presence of oil films.

Springs in the recharge zone of Upper Cretaceous strata contain organic carbon in amounts ranging from 0.4 to 4.4 mg/liter, the average value being 2.0 mg/liter (based on 5 analyses). In the vicinity of Miatli there are two exploratory wells, drilled to the Upper Cretaceous strata, from which flowing water with no signs of oil or gas is obtained (from a depth of 480-600 m). The content of organic carbon in both samples is very small: 1.3 and 1.7 mg/liter.

In southern Dagestan we studied waters from the Khadum beds at the gas fields of Dagestanskie Ogni and Khosh-Menzil. Here all the wells showed gas with the water. All the wells flow spontaneously. We made but few analyses of the waters at the gas deposits, and they are therefore all shown in Table 26.

The very limited factual material we have on waters from the Khadum and Upper Cretaceous strata point to a low content of organic material in these waters, regardless of the zone from which they come, the chemical composition, the presence of gases, etc.

A similar low content of organic material is also found in waters from Lower Cretaceous strata. These waters are encountered in two places: an exploratory well on the Él'dema uplift and at the Khosh-Menzil gas field. In both cases organic carbon is present in amounts less than 6 mg/liter.

Naphthenic acids in the waters at gas deposits are present in amounts ranging from 0.2 to 0.6 mg/liter. Only in three of 11 water samples were they not detected.

Chapter VI

RESULTS OF MICROBIOLOGICAL INVESTIGATIONS

In keeping with the method of field work proposed by M. E. Al'tovskii, three series of samples were collected for bacteriological investigations:

a) from zones where the rocks are exposed in mountains or, what is the same thing, from zones of ground-water recharge;

b) from wells situated near oil and gas deposits;

c) from zones of ground-water discharge, i.e., from districts of mineral and thermal springs.

Thus, bacterial studies were made of waters collected both from shallow depths, at springs emerging in the recharge zone of the water-bearing strata, and from wells, where the aquifers occur at depths ranging up to 1700-1800 m below the surface, and where the water temperatures are as high as 70-85°.

Although the Soviet scientists B. L. Isachenko [1939], A. E. Kriss [1953], L. D. Shturm [1950], and S. I. Kuznetsov [1950] and foreign microbiologists [ZoBell, 1952] have shown that active microflora live at great depths both in the seas and oceans and in ground water in rocks, there is ever present an opinion that deep formational waters, in the absence of faults, must be sterile because of high temperatures and pressures. It has been thought, according to this opinion, that bacterial samples from great depths were collected with insufficient care and that microorganisms were introduced during drilling (with drilling muds and drilling instruments).

ZoBell [1949] studied the relationship of water temperature and hydrostatic pressure to the development of microorganisms in natural environments. To accomplish this, he set up a series of laboratory experiments with bacteria taken from soil and from sea water at depths down to 5000 m. The hydrostatic pressure was varied between 1 and 600 atm in these experiments, the temperature between 20 and 40°C. It was shown that soil bacteria, which normally grow at a temperature of 30°C and at a pressure of 1 atm, are not able to multiply at a pressure of 600 atm; and after 48 hours most of the bacteria die. Bacteria taken from great depths in the ocean (barophilic microflora) thrive under artificial conditions at a pressure of 600 atm, both at 30°C and at 40°C. Marine bacteria taken from the surface layers of mud occupy an intermediate position. Most of them do not survive high pressures, but some varieties are encountered which will thrive at pressures from 400 to 600 atm and at a temperature of 40°C. At lower temperatures the growth of the bacteria is retarded at these same pressures.

ZoBell believes that, in the temperature interval permitting cultures to develop at normal pressure, the retarding effect of high pressures on bacterial growth decreases with increase in temperature.

On the basis of ZoBell's work, it seems to us that microflora introduced into wells from the soil and from air during drilling, being carried down to great depths, should perish from the high temperature and pressure. Specific microlfora that thrive at high pressures, living at great depths, could develop only after long periods of geologic time, by adapting themselves to the conditions of their habitat, that is, to high pressures and temperatures. In studying the bacterial population of ground water we cannot perform experiments at high pressures, and therefore we may form our judgments about the presence of bacteria in deep ground water only by general computations of bacteria under the microscope, with due consideration to the individual physiological groups of bacteria at atmospheric pressure. The computation of total numbers of bacteria was made directly under the microscope by A. S. Rasumov's method [1932] and on glucose-peptone agar in Petri dishes.

The computation of intensity of bacterial development for the different physiological groups was made arbitrarily on a five-point scale; in doing this, the rate at which some groups formed gas was considered, in others, the blackening of the culture medium or the rate of forming films.

Aerobic, saprophytic species of bacteria, able to use easily assimilable compounds of albumin and hydrocarbon character, were counted on glucose-peptone agar.

Putrefying bacteria, which decompose albumin substances with the formation of hydrogen sulfide and hydrogen, were determined in a medium containing lead acetate.

Denitrifying bacteria, which reduce nitrates to gaseous nitrogen, were isolated on two media: bacteria that use ready-made organic compounds as a source of carbon were grown on a medium with calcium citrate; bacteria that do not need ready-made organic material for their carbon and that use, as a source of energy, the oxidation of thiosulfate salts or sulfur to sulfuric acid were grown on Liske medium with hyposulfite.

Desulfurizing bacteria, which reduce sulfates to hydrogen sulfide in the presence of various organic substances, were isolated on Shturm medium with calcium lactate.

Desulfurizing autotrophic species were isolated in an atmosphere of hydrogen according to a method proposed by Yu. I. Sorokin [1952]. Thionic acid bacteria, which oxidize thiosulfates or hydrogen sulfide to sulfuric acid, were isolated on Natanson medium. The intensity of their development was computed by the quantity of hydrosulfite used up.

Bacteria that decompose cellulose in an anaerobic environment, with the formation of hydrogen, carbon dioxide, and methane, were grown on Omelyanskii medium.

Bacteria that form methane by decomposing acetates, ethyl alcohol, or gaseous hydrogen and carbon dioxide were isolated on Barker medium.

Bacteria that oxidize hydrogen or hydrocarbons were segregated in liquid Muntz medium in an atmosphere of analyzing gas. Putrefying, denitrifying, cellulose, and methane-forming cultures of bacteria were isolated on an agar medium.

In order to get some idea of the consistent development of any particular physiological group of bacteria in an aquifer at some point being investigated, two analyses of the water were made at each of 24 points. Parallel samples were collected a year or two later. A comparison of the data has shown that at points where the intensity of bacterial growth corresponded to one or two on the scale, the second analyses frequently failed to agree. Where bacteria failed to grow or where the scale value was 3-5, the second samples almost always gave similar results, although a few showed some change. For example, instead of 3, the second value was 5; or instead of 4, a 2 was recorded.

Obviously, we may consider the development of bacteria to be consistent when the scale value becomes 3 or higher. On repeating analyses on glucose-peptone agar, the order of value for bacterial quantity did not change, except for rare exceptions.

For single bacterial analyses we collected two parallel samples in flasks with ground-glass stoppers, which had been previously sterilized in an autoclave. The water from one flask was analyzed in the field laboratory; the water from the second was analyzed at the permanent station. As for chemical analyses, water was obtained from wells by taking it from the test cocks by filling the flasks from a stream of water, the only difference being that the samples were collected in a sterile vessel for the bacterial analyses.

The collection of water samples for microbiological study from pressurized pipes was made at least 1-2 hours after the resumption of pumping where pumping was intermittent, and, even then, only if the pipe was full of water.

In those cases where samples could not be collected from a stream of water from a spring, the samples were obtained by submerging a flask in the water.

In the field laboratory the samples were prepared for direct computation of bacteria under the microscope and for quantitative evaluation on agar plates; and, in addition, reserve cultures of denitrifying, desulfurizing, putrefying, methane-forming, cellulose, and thionic acid bacteria were isolated.

The samples were analyzed within 24 hours of the time they were collected. More detailed studies of the bacterial microflora were made at the permanent laboratory in Moscow.

In 1954 we analyzed 69 samples in the Grozny region, and in 1955, 57 samples in the Dagestan ASSR. In examining the results obtained from studying specific microflora in relation to physicochemical conditions in the environment, we found the results from the Grozny region to be similar to those for the Dagestan ASSR. Therefore, in this book we have presented detailed data on abundance of microflora for the three separate physicochemical zones only for the Grozny region. For the Dagestan ASSR, we have shown data on abundance of bacteria found in the Chokrak water-bearing horizon (series "B").

THE ZONE OF EXPOSED STRATA IN THE CHERNYE MOUNTAINS

The first series of samples was collected in the recharge zone for the aquifers in the Chernye Mountains.

The samples were obtained at springs situated in villages or along roads passing between villages.

Chemically the springs in the recharge zone (data in Table 27) may be described as slightly mineralized, with low organic content.

Mineral nitrogen is present in nitrates and ammonium salts. Oxygen is present in quantities ranging from several milligrams down to hundredths of a milligram per liter. Hydrogen sulfide is absent. The oxidation-reduction potential ranges from 30 to 17.8 for values of rH_2. The water temperature ranges from 12.5 to 21°C. Most of the springs have low discharges.

Let us go on to a description of the bacterial composition of the water. In total bacterial content, as seen from Figure 5, one milliliter of water contains from 59,000 to 200,000 bacteria. It should be noted that springs with bacterial contents from 59,000 to 83,000 bacteria per milliliter of water (Buinaksk, spring No. 1 in Benoi) were better tapped than the others. It is likely that the lower bacterial content in these springs is due to this fact.

TABLE 27

Chemical Composition of Water According to Zones

Sample locality	t°	Total minerali-zation, mg/liter	rH_2	SO_4^{2-}, mg/l	H_2S, mg/l	O_2, mg/l	NH_4, mg/l	NO_3^-, mg/l
Springs in recharge zone	12.5–21	60–1567	30.4–17.8	20–543	0–0.2	5.62–0.59	1.75–0.8	75–0.9
Ground water at oil deposits	23.5–89	1000–5000	17.8–9.7	0–430	0–40	0	18.9–0.7	0
Mineral springs (discharge zone)	12.5–87	814–6236	15.2–7.6	29–1267	4–57	0	10–0.9	0

In untapped springs (Buinaksk, spring No. 2, Kydyrovo, Benoi, a spring beside the road and a well in Benoi) the bacterial content ranges from 126,000 to 920,000 bacterial cells per milliliter of water.

In computing the total number of saprophytic aerobic bacteria in springs in the first physicochemical zone (recharge zone), we found great numbers of bacterial colonies per milliliter of water (from 660 to 15,000).

According to A. S. Razumov [1932], the ratio of total number of bacteria (computed directly under the microscope) to saprophytes characterizes the composition of bacterial microflora. In sewage, where specific saprophytic microflora are developed, this ratio is equal to unity; i.e., all the microflora counted directly under the microscope are saprophytic, as well as all those planted on meat-peptone agar. In water from surface streams the ratio of total bacteria to saprophytes ranges from tenths down to hundredths to one, since only some of the bacteria grow on meat-peptone agar culture. In water that possesses specific microflora, this ratio reaches a thousand and even a hundred thousand to one, since most of the microflora do not grow on meat-peptone agar. In the springs of the described zones, the ratio of total bacteria to saprophytes is everywhere small: it ranges from several tens up to a hundred to one, i.e., near the value found in surface streams.

Going on to a description of the bacterial composition according to physiological groups, we may note that denitrifying, cellulose-destroying, putrefying, and hydrogen-oxidizing bacteria are abundant in all samples from the recharge zone of the aquifers.

Desulfurizing and methane-forming bacteria are not abundant and do not have any fundamental effect on the gases or on the chemical composition of the water in this zone.

Thionic acid and methane-oxidizing bacteria are not present at all.

THE GROUND-WATER ZONE IN THE VICINITY OF OIL DEPOSITS

Samples of water from this zone were collected from wells drilled in districts of oil deposits, from depths ranging from 210 to 2123 m. The water temperature ranged from 23.5 to 89°. The total mineralization in most samples was 1-5 g/liter, but in individual samples ranged from 0.6 to 39.0 g/liter. The sulfate content fluctuated widely, reaching as much as 430 mg/liter in some samples. On the other hand, some samples proved to have no sulfate at all. The hydrogen-sulfide content ranged from 0 to 40 mg/liter. No oxygen was detected. The oxidation-reduction potential ranged from 17.8 to 9.7 for values of rH_2.

The content of organic carbon ranges from several milligrams to 70 milligrams per liter. Nitrogen in the form of nitrates was missing from all samples; ammoniacal nitrogen was present in amounts ranging from tenths of a milligram to several milligrams per liter (see Table 27).

In considering the data on total bacterial content in the water, as determined directly by computation under the microscope (see Fig. 5), we see that one milliliter of water contains from 25,000 to 9,600,000 bacterial cells, the waters with high temperatures (from 89 down to 55°) having fewer bacteria (25,100 to 83,700) than waters with low temperatures (from 50 down to 20°), where the number of bacteria ranges from 105,000 to 9,600,000 per milliliter of water.

In computing saprophytic bacteria on glucose-peptone agar, 10 to 400 bacteria per milliliter of water were identified.

On the average, the ratio of total bacteria to saprophytes in water of this zone is characterized by a high value, on the order of several thousand to one, a fact that points to the development of specific bacterial species in this water.

In considering the average abundance of bacteria in the different physiological groups, we see that, in this zone, as in the recharge zone, denitrifying and cellulose-destroying bacteria are abundant. On the average, their abundance may be evaluated at 4 on the scale. If we examine the bacterial abundance for individual deposits, then we should note that samples collected in the Oktyabr'skoe field (waters with high temperatures), bacteria that decompose cellulose (with the formation of gas) were not present. Bacteria that form methane through the reduction of carbon dioxide by gaseous hydrogen were found in almost all the samples we analyzed, even in samples of hot water collected at the Oktyabr'skoe field, but their abundance in the zone as a whole was placed at 2 on the scale. The formation of methane by bacterial decomposition of acetates was slight, although it was more energetic than in water from the ground-water recharge zone.

Bacteria that oxidize hydrocarbons and hydrogen are aerobic bacteria; however, there are some who believe that, under certain circumstances, when oxidation-reduction conditions are suitable, such bacteria may develop in an anaerobic environment. The results of analyses on cumulative cultures of hydrogen- and carbohydrate-oxidizing bacteria have shown that only heptane-oxidizing bacteria were found in large numbers in the samples collected in the vicinity of oil deposits (Oktyabr'skoe, Starogrozny, Tashkala, and Novogrozny). Methane- and hydrogen-oxidizing bacteria were found in small numbers at the Malgobek field.[*]

---

[*] The isolation of cultures of bacteria that oxidize hydrogen and hydrocarbons was effected in the laboratory of G. A. Mogilevskii. The cultures were isolated a long time after collection of the sample. It is possible that clearer results might be obtained on freshly collected samples.

TABLE 28

Abundance of Desulfurizing Bacteria in Relation to the Presence of Organic Material and to the Temperature of the Water

Sample No.	Sample locality	Designation of aquifer	Water temp., °C	SO_4^{2-}, mg/liter	H_2S, mg/liter	C (acc. to Datsko), mg/liter	Abundance of bacteria, scale value
		Starogrozny oil field					
	Starogrozny field:						
32	Well 41/2	II bed in thrust plate	23	6		30.3	5
30	Well 5/52	ec	27.5	7		70.6	3
34	Gunyushki, well 1	VII-VIII	76	438		3.1	0
41	Solenaya balka, well 5/92	XII in thrust plate	52	2	10	13.5	2
37	2nd subsidiary firm, well 8/0	XII "	44	188	0	6.5	0
40	Tashkala, well 8/86	XII beneath thrust plate	48	6	4	16.2	4
	Tashkala:						
39	Well 8	XII "	66.5	67	15	25.6	0
38	Well 24	XII "	68.5	205	10	13.0	0
36	Well 145	XII "	54	337	0	12	0
31	Starogrozny field, well 22/977	XVI beneath thrust plate	44	0		45.2	5
		Gas fields of the Dagestan ASSR					
	Khosh-Menzil:						
29/38	Well 18	Khosh-Menzil	24	0	0	4.7	0
30/39	Well 21	The same	39.5	0	0	3.5	0
	Dagestanskie Ogni:						
34/46	Well 37	Khadum	39	0	0	2.8	0
35/47	Well 17	Maikop	39	0	0	3.3	0
36/50	Well 19	Upper Cretaceous	41	0	0	4.2	0
	Duzlak:						
46/66	Well 53	Lower Cretaceous	63	0	0	3.6	0
47/67	Well 46	The same	34	0	0	60.0	0
48/69	Well 21	Khadum	32.5	0	0	18.8	0

Karagan

Chokrak

The abundance of desulfurizing bacteria has been shown, in general, by a scale value of 3; this abundance in the water is due to the presence of organic material and sulfates and to proper temperatures in the water.

We investigated the waters from seven oil fields in the Grozny region and from four in the Dagestan ASSR. The results were everywhere the same. Somewhat different results were obtained for waters from gas fields in the Dagestan ASSR. We shall confine ourselves to an examination of the results obtained for the Starogrozny field, from waters in the Karagan and Chokrak strata, and for gas fields in Dagestan, where samples were collected from Cretaceous rocks and the Khadum horizon.

It may be seen from Table 28 that waters taken from aquifers in the Karagan strata at the Starogrozny field, waters which contain 30-70 mg/liter organic material (wells 41/2 and 5/52), show active growth of de-sulfurizing bacteria at temperatures of 23-27°. The sulfate content in these waters ranges from 6 to 7 mg/liter, whereas desulfurizing bacteria do not grow in the water in well 1 at Gunyushki, where the organic content is 3.1 mg/liter and the water temperature is 76°, even though the sulfate content reaches 438 mg/liter.

Of two samples collected from bed XII of the Chokrak strata (in the thrust plate), desulfurizing bacteria were identified in the sample from Solenaya balka (well 5/92), where the organic content amounted to 13.5 mg/liter. In the sample taken at Tashkala (well 8/0), where the organic content was less (6.5 mg/liter), these bacteria were not found, as a result of which the sulfate content in this water was considerably higher (188 mg/liter) than at Solenaya balka (2 mg/liter). Of four samples collected from bed XII below the thrust plate, where the organic content was high (12-25 mg/liter), desulfurizing bacteria were detected in but one sample (well 8/86), where the water temperature was 48°. In the remaining samples the water temperature was higher (54-68°C), a fact that restricted the growth of these bacteria.

In considering the results of analyses of water samples collected from Cretaceous and Khadum strata at gas fields in Dagestan, we find a different picture. Desulfurizing bacteria were not found in a single sample. Here the limiting factor in bacterial growth, as seen from Table 28, is the absence of sulfates in the water.

Thus, on the basis of the work done, it may be stated that, in the waters from the regions we investigated, the abundance of desulfurizing bacteria is closely connected with the presence of organic material, the content of sulfates, and the temperature of the water. And where desulfurizing bacteria are present the quantity of sulfates is markedly lowered in the water.

It should be noted that putrefying bacteria are not present in the vicinity of oil deposits, but, on the contrary, a different group of bacteria appears, varieties that oxidize thiosulfates to sulfuric acid; these are not found in the zone of ground-water recharge. Their average abundance over-all is 2 on the scale. If we consider their abundance in individual deposits, we must point out that they are most abundant at the Starogrozny field.

THE DISCHARGE ZONE IN REGIONS OF EMERGENCE OF MINERAL AND THERMAL WATERS

The third series of samples was collected from mineral and thermal springs. Most of these springs are used by the local inhabitants for medicinal purposes.

Chemically this group of springs is distinguished from the springs of the first group by higher mineralization: 0.8-6.2 g/liter (see Table 27). The oxidation-reduction potential ranges from 15 to 7.6 for values of rH_2. All the springs contain hydrogen sulfide in quantities ranging from 4 to 57 mg/liter. The sulfate content ranges from 28.8 to 1267 mg/liter. Nitrates are not present in most springs. Ammoniacal nitrogen is present in amounts of 0.9 to 10 mg/liter. Organic carbon ranges from 1.8 to 16 mg/liter. The water temperature at most springs lies somewhere between 35 and 87°C, although some springs have temperatures between 12.5 and 25°C.

According to total bacterial content in the waters of the mineral springs, as in the waters from the vicinity of oil deposits, all samples may be divided into those from thermal springs (with temperatures between 63 and 86°C) and those from low-temperature springs (temperatures from 52 down to 12°C). The waters of the first group contain bacteria in numbers that range from 8600 to 39,100 per milliliter of water; in the second group the range is from 144,000 to 3,100,000 per milliliter of water.

The number of saprophytic bacteria grown on glucose-peptone agar ranges from individuals to hundreds per milliliter of water.

Exceptions to this are found in samples collected at Miatli (iron spring) and at Goryachii Klyuch (spring No. 4), where the content of saprophytic bacteria reaches 5000-9000 per milliliter of water. It is possible that these springs have been contaminated, since they are used by the inhabitants for medicinal purposes.

In most of the samples analyzed, the ratio of total bacteria count to saprophytic forms is greater than a thousand to one, a fact which indicates that specific microflora are also developed in the mineral springs.

Denitrifying bacteria grow equally well in a medium with calcium citrate and in a medium with hyposulfite; this is true of all samples except for those where the water temperature has reached 83-87°C. Bacteria that decompose cellulose and that produce methane through the reduction of carbon dioxide by gaseous hydrogen are widespread in the waters of these mineral springs, but they are less abundant here than in the vicinity of oil deposits, as may be seen from Figure 6.

Cellulose-destroying bacteria are the most abundant. They were identified in almost all samples, except for those having high temperatures. Methane-forming bacteria, on the other hand, are abundant in waters with high temperatures.

Bacteria that destroy acetates with the formation of methane were found only in some springs.

In contrast to bacteria that produce methane, bacteria that oxidize hydrocarbons were found to be almost completely absent in the waters of these mineral springs, a fact apparently to be explained by the presence of dissolved hydrogen sulfide in the springs.

Hydrogen-oxidizing bacteria were identified in waters from a number of springs, but their average abundance was low.

Desulfurizing bacteria are widespread in the waters of the mineral springs. Of 21 samples from the Grozny region, desulfurizing bacteria could not be found in only eight, and four of these samples were from thermal springs with temperatures in the range 75-83°C. The average abundance of these bacteria has been placed at 2-3 on the scale. According to content of organic carbon, the waters of these mineral springs differ sharply from waters at oil deposits. The quantity of organic carbon ranges, on the average, from 2 to 5.6 mg/liter. Despite this small content of organic carbon, desulfurizing bacteria are rather abundant in the water. This fact may be explained only if the source of organic material or molecular hydrogen for the development of the bacteria come from somewhere outside the water associated with oil deposits.

Despite the abundance of desulfurizing bacteria, the springs of this group contain large quantities of sulfates (see Table 27), as much as 1200 mg/liter. Apparently the accumulation of sulfates in the water is promoted by the active growth of thionic bacteria, which oxidize hydrogen sulfide and thiosulfates to sulfates.

As may be seen from Figure 6, thionic acid bacteria are most abundant in the discharge zone of ground water. The average abundance is 4 on the scale.

DAGESTAN ASSR

Seven samples of water were collected from wells at oil deposits and beyond the oil boundary from the petroliferous Chokrak strata (horizon "C") at the Ternair and Makhachkala fields in northern Dagestan. The data on these samples are shown in Table 29.

The first part of the table includes results of analyzes, for comparative purposes, of water collected from Chokrak strata also, but from springs discharging in the recharge zone of the aquifers (seven kilometers from Makhachkala). As seen from Table 29, the total bacterial count, as computed directly under the microscope, reaches 659,500 bacteria per milliliter of water in the recharge zone of the aquifers.

In the vicinity of oil deposits, at high temperatures of 45-60°C, the total number of bacteria in the water ranges from 20,000 to 70,000 per milliliter of water. An exception is found in well 76 at the Makhachkala field, where the water temperature is 25°C and the organic content is high (8.2 mg/liter of organic carbon). The total bacterial count in the water here is as much as 844,000 per milliliter of water.

Saprophytic and putrefying bacteria are abundant only in the recharge zone of the aquifers, but denitrifying, cellulose-destroying, and methane-forming varieties are equally abundant in the recharge zone and in the vicinity of oil deposits.

T A B L E   29

Change in Bacterial Composition of Water in the Zone of Stagnant Water along the Dip of an Aquifer (Horizon "C") of Chokrak Strata

Sample No.	Sample locality	Water temperature, °C	Depth, m	Presence of oil	Total bacterial count, thousands per ml of water	Content of saprophytes in scaled values	Abundance of bacteria in scaled values						SO_4^{2-}, mg/liter	H_2S, mg/liter	C_{org} (acc. to Datsko), mg/liter
							Putrefying	Denitrifying	Cellulose	Methane-forming	Desulfurizing	Thionic			
12/14	Recharge zone, springs (along the road from Makhachkala to Talgi)	18.3	—	—	659.5	5200	2	3	3	5	1	—	295	0	0.9
	Ternair:														
8/10	Well 10	48	1646-48	+	42.3	465	—	3	3	—	4	—	12	0.3	10.5
7/9	Well 20	61	1661-65	—	32.2	2	—	2	3	—	—	3	1215	0.1	0.9
6/8	Well 125	60	1656-58	—	40.6	3	—	—	3	5	—	2	1193	0	2.1
	Makhachkala:														
3	Well 27	> 50	1505-9	—	70	151	—	—	1	—	—	3	570	0	0.5
2	Well 76	25	1479-81	+	844	144	—	5	—	5	5	1	1	—	8.2
1	Well 75	> 50	1453	+	63.6	113	—	5	3	—	—	—	1	—	4.8
53/6	Well 30	45	1217-30	—	20.7	—	—	1	5	5	—	1	598	0.1	1.0

In the aquifer recharge zone, where the content of organic carbon is 0.9 mg/liter, desulfurizing bacteria have an abundance represented by 1 on the scale. At the Ternair field, where the content of organic carbon is greater (10.5 mg/liter), in a sample taken from oil well 10, the abundance of these bacteria is 4 on the scale; the sulfate content drops from 295 mg/liter in the recharge zone to 12 mg/liter at the oil well, and hydrogen sulfide appears at the latter locality (0.3 mg/liter).

In wells 20 and 125, beyond the oil boundary, the content of organic carbon drops to 1-2 mg/liter, the water temperature rises to 60°C, desulfurizing bacteria do not grow, and the content of sulfates in the water increases to 1200 mg/liter. A similar picture is found at the Makhachkala oil field. In well 27, beyond the oil boundary, the content of organic carbon is 0.5 mg/liter and desulfurizing bacteria are absent. In wells 76 and 75, situated at the oil boundary, the abundance of bacteria is antithetic. At well 76, where the content of organic carbon is 8.2 mg/liter and the temperature is 25°C, desulfurizing bacteria are abundant and the sulfate content in the water again falls to 1 mg/liter. In well 75, despite a comparatively high content of organic carbon (4.8 mg/liter), desulfurizing bacteria are absent. Apparently this relationship is associated with an increase in water temperature (above 50°C) and an insufficient quantity of sulfates in the water (1 mg/liter).

In well 30, beyond the oil boundary, where the content of organic carbon is 1 mg/liter and the water temperature is low (down to 45°C), desulfurizing bacteria are not present in an environment containing organic material, and the quantity of sulfates in the water increases to 598 mg/liter. It should be noted that we were able to isolated desulfurizing bacteria from this sample in a medium with gaseous hydrogen, but their abundance was low (scale 1). It is possible that the presence of small quantities of hydrogen sulfide (0.1 mg/liter) was associated with the growth of this group of bacteria.

Concerning the abundance of thionic acid bacteria, it may be stated that they began to develop extensively in aquifers after desulfurizing bacteria (see Table 29).

Thus, the data we have obtained on the abundance of bacterial microflora in certain aquifers agree fully with the data on changes in bacterial forms in zones as a whole obtained for the Grozny region.

SUMMARY

From the results of our microbiological investigations, we may draw the following conclusions.

1. In the recharge zone of aquifers the total bacterial count in the water ranges from several hundred thousand to several millions per milliliter of water.

2. In the vicinity of oil deposits and at discharge zones of the aquifers, the total bacterial count in the water is a function of the temperature. In hot waters (55-89°C) the count is in the thousands, but in waters with temperatures of 20 to 50°C, the number of bacteria is in the hundred thousands and millions.

The aerobic saprophytic group of bacteria, which use glucose and peptone as a source of organic material, are specific microflora for the recharge zone.

Putrefying bacteria, which decompose albumin substances with the formation of hydrogen and hydrogen sulfide, and hydrogen-oxygen bacteria are abundant in the recharge zone of the aquifers, are almost absent in the vicinity of oil deposits, and are but sparsely present in the discharge zone.

3. Denitrifying bacteria, which decompose cellulose in an anaerobic environment and which form methane through the reduction of carbon dioxide by hydrogen, are abundant in all the zones we investigated.

4. Desulfurizing and heptane-oxidizing bacteria are sparsely present in the recharge zone and abundant in water associated with oil deposits. They are less abundant in the discharge zone of the aquifers.

5. In waters associated with oil deposits, where the waters contain sulfates, the abundance of desulfurizing bacteria depends on the presence of organic material and on the temperature of the water. Where sulfates are absent, desulfurizing bacteria do not develop.

6. Thionic acid bacteria are not present in the recharge zone of aquifers, appear in the vicinity of oil deposits, and are most abundant in the discharge zone of the aquifers.

BIBLIOGRAPHY

G. V. Abikh, "Oil deposits in the southwestern Caucasus," Byull. MOIP, 11 (1867).

M. V. Abramovich, "Changes in the properties of oil in oil starta in connection with the mode of occurrence," Trudy Geol. In-ta im. Gubkina, Azer. FAN, 19 (1939).

M. V. Abramovich, "The relationship between specific gravity of oil and the mode of occurrence within several strata in the Bibi-Éibat area (Stalino Oil Trust)," Trudy Geol. In-ta im. Gubkina, 20, Azer. FAN (1941).

T. P. Afanas'ev, "Chemical zoning in the ground water of the middle Volga region," Doklady Akad. Nauk SSSR, 56, 6 (1947).

S. N. Alekseichik, "The possibility of oil formation in continental formations," Neftyanoe Khozyaistvo, 12 (1946).

M. E Al'tovskii, The Problem of Chemical Evolution in Ground Water. Questions on Hydrogeology and Engineering Geology [in Russian] Collection No. 13 (All-Union Scientific Research Institute of Hydrogeology and Engineering Geology, 1950).

N. I. Andrusov, "The origin and occurrence of oil," Tr. Bak. Otd. Imper. Russk. Tekhn. Ob-va, Book 1-2 (1908).

A. D. Arkhangel'skii, "The antiquity of hydrogen-sulfide contamination in the marine basins of the Crimea-Caucasus region and the probable relationship between this phenomenon and oil formation," Neftyanoe Khozyaistvo, 4 (1926).

A. D. Arkhangel'skii, Conditions of Formation of Oil in the Northern Caucasus [in Russian] (Council for the Petroleum Industry Press, 1927).

A. D. Arkhangel'skii and É. S. Zalmanzon, "Development environment of waters in the Grozny oil fields," Trudy Neft. Geol. Razv. In-ta, Collection of papers for 1930-31 and 1932.

K. B. Ashirov, "The teachings of I. M. Gubkin concerning the conditions under which oil originates and under which oil and gas deposits are formed," Neftyanoe Khozyaistvo, 4 (1954).

P. P. Avdusin, "Sandy rocks in the Visean sequence of Vtoroi Baku as oil reservoirs," Doklady Akad. Nauk SSSR, 42, 1 (1944).

P. Ya. Avrov et al., The Geologic Structure and Oil Fields of the Emba Region [in Russian] (United Scientific and Technical Press, 1935).

A. A. Bakirov, The Geologic Structure and Oil Potential in the Paleozoic Strata of the Central Russian Structural Basin [in Russian] (State Scientific and Technical Press of the Petroleum and Mineral-Fuel Industry, 1948).

A. A. Bakirov, "Serious, fundamental errors of Professor N. A. Kudryatsev concerning theories of the origin of oil," Neftyanoe Khozyaistvo, 9 (1951).

A. A. Bakirov, "The present status and urgent tasks of investigations on oil migration in the earth's crust," Sovetskaya geologiya, collection 47 (1955).

A. A. Bakirov and F. O. Mirchink, "Some questions concerning the theory of geotectonic development of large structural elements of the earth's crust in relation to the study of their oil and gas potential," Neftyanoe Khozyaistvo, 9 (1951).

N. D. Balukhovskii, "The course of further development in the teachings of Academician I. M. Gubkin concerning the origin of oil," Neftyanoe Khozyaistvo, 11 (1952).

V. P. Baturin, "The origin of oil," Byull. MOIP, Novaya Seriya, Otdel Geologii, 20, 1-2 (1945).

V. V. Belousov, "The origin of natural gases of the methane group," Neftyanoe Khozyaistvo, 12 (1935).

E. Berl, "The origin of oil," American Institute of Mining and Metallurgical Engineering Technology, 920, Petroleum Technology (1937).

I. V. Bochkov, V. G. Vasil'ev, and others, The Ural-Volga Oil District [in Russian] (State Scientific and Technical Press of the Petroleum and Mineral-Fuel Industry, 1941).

I. O. Brod, The Problem of the Origin of Oil and Gas, Collection: "Prirodnye gazy" (Natural Gases), 4-5 (1932).

I. O. Brod, "A classification of oil and gas deposits in relation to the fluids in natural reservoirs," Doklady Akad. Nauk SSSR, 53, 1 (1946).

I. O. Brod, "The principles of subdividing oil and gas districts," Doklady Akad. Nauk SSSR, 53, 5 (1946).

I. O. Brod, "Basic conditions for oil accumulation," Doklady Akad. Nauk SSSR, 57, 6 (1947).

I. O. Brod, "The classification of oil deposits according to their forms (as illustrated by fields in the northeastern Caucasus)," Transactions of the 17th session of the International Geological Congress, 4 (1947).

I. O. Brod, "The migration of fluidal hydrocarbons in the earth's crust," Izv. AN SSSR, Seriya Geologii, 4 (1947).

I. O. Brod, "The subdivision of oil and gas districts," Trudy Mosk. Neft. In-ta, 5 (1947).

I. O. Brod, Basic Questions on Oil and Gas Formation. The Origin of Oil and Natural Gas [in Russian] (Information Office for Technology and Economics. Press of the Central Scientific Research Institute for Mechanization and Labor Organization in the Petroleum Industry, 1947).

I. O. Brod, "Oil and gas formation in the light of current data," Nov. Neft. Tekhn. TsIMT, Seriya Geologii, 11 (1948).

I. O. Brod, Oil and Gas Deposits [in Russian] (State Scientific and Technical Press of the Petroleum and Mineral-Fuel Industry, 1954).

I. O. Brod and N. A. Eremenko, "The migration of oil and gas and the classification of migration processes," Vestnik Mosk. Gos. Un-ta, 3 (1947).

I. O. Brod and N. A. Eremenko, "Discussions on the origin of oil and the formation of oil deposits," Neftyanoe Khozyaistvo, 5 (1951).

I. O. Brod and N. A. Eremenko, Principles of the Geology of Oil and Gas [in Russian] (Moscow State University Press, 1953).

I. O. Brod and E. F. Frolov, Prospecting and Exploring for Oil and Gas Deposits [in Russian] (State Scientific and Technical Press of the Petroleum and Mineral-Fuel Industry, 1950).

I. O. Brod and V. G. Levinson, The Origin of Oil and of Oil and Gas Accumulation [in Russian] (State Scientific and Technical Press of the Petroleum and Mineral-Fuel Industry, 1955).

I. O. Brod, V. G. Levinson, and N. A. Eremenko, Geologic Conditions for the Formation of Bitumens [in Russian] (Moscow State University Press, 1948).

S. V. Bruevich, "The chemical composition of bottom-mud solutions in the Caspian Sea," Collection "Gidro-khimicheskie materialy" (Hydrochemical data), 13, Academy of Sciences, USSR (1947).

S. V. Bruevich and E. G. Vinogradova, "Accumulation of biogenetic elements in bottom-mud solutions in the northern Caspian," Doklady Akad. Nauk SSSR, 27, 6 (1950).

N. A. Buneev, P. A. Kryukov, and E. B. Rengarten, "Experiments on squeezing solutions from sedimentary rocks," Doklady Akad. Nauk SSSR, 57, 7 (1946).

N. I. Butorin and Z. P. Buks, "Naphthenic acids in formational waters of the Starogrozny region," Groznenskii Neftyanik, 5-6 (1935).

N. I. Buyalov, "The Emba oil district," Transactions of the 17th session of the International Geological Congress, 4, 1940.

M. M. Charygin, The Physical-Geographic and Geologic Conditions for the Accumulation of Source Material for Bitumen Formation [in Russian] (Technical Information Collection on Geology, 1947).

A. A. Cherpeninkov, Hydrogen in Natural Gases [in Russian] (Collection of Papers Commemorating the Fiftieth Anniversary Jubilee of Academician V. I. Vernadskii, 1936).

P. Chirvinskii, "A Survey of the principal hypotheses concerning the origin of oil," Neftyanoe Khozyaistvo, 7 (1925).

Craig, see also Kreg.

G. Craig, The Geochemistry of the Stable Isotopes of Carbon. Isotopes and Geology [translated from English] (Foreign Literature Press, 1954).

V. G. Datsko, "Soluble organic material and its accumulation in sedimentary rocks," Doklady Akad. Nauk SSSR, 59, 3 (1948).

V. G. Datsko, "A method of determining organic carbon in natural waters," Doklady Akad. Nauk SSSR, 73, 2 (1950).

V. G. Datsko, "The vertical distribution of organic material in the Black Sea," Doklady Akad. Nauk SSSR, 77, 6 (1951).

A. F. Dobryanskii, "The mutual relationships of oils, asphalts, and sapropelites," Azv. Azer. Filial Akad. Nauk SSSR, 5 (1944).

A. F. Dobryanskii, The Geochemistry of Oil [in Russian] (State Scientific and Technical Press of the Petroleum and Mineral-Fuel Industry, 1948).

A. F. Dobryanskii, "A reply to the paper of A. Ya. Krems," Neftyanoe Khozyaistvo, 1 (1951).

N. D. Elin, Oil Fields of the Terek Anticlinorium. Collection of Data for the 17th International Geological Congress. Oil Fields of Eastern Ciscaucasia [in Russian] (State All-Union Association of the Grozny Oil and Gas Industry Press, 1937).

N. A. Eremenko, "Investigations in the developmental conditions of oil deposits from considerations of changes in the properties of the oil," Nov. Neft. Tekhn. TsIMTnefti, Seriya Geologii, 6 (1948).

N. A. Eremenko and M. S. Bezhaev, An Investigation of Waters at Oil Fields (as Exemplified in Dagestan) [in Russian] (State Scientific and Technical Press of the Petroleum and Mineral-Fuel Industry, 1956).

M. V. Fedorov, Microbiology [in Russian] (State Press of Agricultural Literature, 1955).

S. F. Fedorov, Oil Fields in the Soviet Union [in Russian] (State Scientific and Technical Press, 1939).

S. F. Fedorov, "The prediction of oil occurrences," Doklady Akad. Nauk SSSR, Nov. Seriya, 28, 1 (1940).

A. V. Frost, "Clays as catalysts in the formation of oil," Trudy. Institut Geologicheskikh Nauk, Akad. Nauk SSSR, 11, (1944).

A. V. Frost, "The role of clay in the transformation of oil in the earth's crust," "Uspekhi Khimii," 6 (1945).

A. V. Frost, The Role of Clay in the Formation of Oil and Gas [in Russian] Collection: The Origin of Oil and Natural Gas (Information Office for Technology and Economics. Press of the Central Scientific Research Institute for Mechanization and Labor Organization in the Petroleum Industry, 1947).

A. V. Frost and L. K. Osnitskaya, On the origin of Oil [in Russian] Collection: In Memory of Academician I. M. Gubkin (Acad. Sci. USSR Press, 1951).

G. Gefer, "The origin of oil," ANKh, 10-11 (11-12)(1922).

V. A. Geitling and A. F. Frost, "The role of clays in the formation of oil in the earth's crust," Neftyanoe Khozyaistvo, 11-12 (1945).

F. F. German, "The formation of oil deposits in the Western Turkmen lowland," Neftyanoe Khozyaistvo, 7 (1951).

T. L. Ginzburg-Karagicheva, "Microbiological investigations on the sulfur-saline waters of Apsheron," ANKh, 6-7 (1926).

T. L. Ginzburg-Karagicheva, "Problems concerning the microbiology of oil," Byull. MOIP, Otdel. Geologii, 11, 1 (1933).

T. L. Ginzburg-Karagicheva, Essays on the Microbiology of Oil [in Russian] (United Scientific and Technical Press, 1936).

T. L. Ginzburg-Karagicheva, "Biogenetic factors in the origin of oil and natural gas," Transactions of the 17th session of the International Geological Congress, 4, Gosgeolizdat (1940).

T. L. Ginzburg-Karagicheva, "The origin of microflora in oil- and gas-bearing strata," Sovetskaya Geologiya, 13 (1947).

T. L. Ginzburg-Karagicheva, The Origin of Living Microflora in Oil- and Gas-Bearing Strata. The Origin of Oil and Natural Gas [in Russian] (Information Office for Technology and Economics. Press of the Central Scientific Research Institute for Mechanization and Labor Organization in the Petroleum Industry, 1947).

T. L. Ginzburg-Karagicheva, "The transformation of organic material in an anaerobic environment by bacteria in petroliferous strata and the investigation of oils of various geologic ages," Trudy Mosk. Filiala VNIGRI, 3 (1953).

N. G. Golovanov, I. V. Brovchinskii, and Z. T. Dem'yanova, "Waxes and their uses," Priroda, 8 (1954).

V. A. Gorin, "The role of faulting in the distribution of oil in the strata of the Apsheron Peninsula," ANKh, 8 (1940).

I. M. Gubkin, "The Grozny oil district," Neftyanoe i Slantsevoe Khozyaistvo, 4-8 (1920).

I. M. Gubkin, "Fundamental problems in developing and exploring for oil deposits in the Novogrozny and Maikop regions," Neftyanoe Khozyaistvo, 11 (1930).

I. M. Gubkin, The Tectonics of the Southeastern Part of the Caucasus in Connection with the Oil Potential of this Region [in Russian] (United Scientific and Technical Press, 1934).

I. M. Gubkin, The Science of Petroleum [in Russian] (United Scientific and Technical Press, 1937).

I. M. Gubkin, The Ural-Volga Oil District [in Russian] (Acad. Sci. USSR Press, 1940).

I. M. Gubkin, "On the problem of the origin of oil deposits in the Northern Caucasus," Transactions of the 17th session of the International Geological Congress, 4 (1940).

I. M. Gubkin, "Mud volcanoes in the Soviet Union and their relations to the occurrence of oil," Transactions of the 17th session of the International Geological Congress, 4 (1940).

L. A. Gulyaeva, "The microelements in the oils and bitumens of the Ural-Volga region," Trudy, Institut Goryuchikh Iskopaemykh, Akademii Nauk SSSR, 2 (1944).

N. K. Ignatovich, "The systematic distribution and development of ground water," Doklady Akad. Nauk SSSR, 44, 3 (1944).

N. K. Ignatovich, "Hydrogeological environment for the formation and preservation of oil deposits," Doklady Akad. Nauk SSSR, 46, 5 (1945).

N. K. Ignatovich, "The hydrogeological classification of geostructural elements," Doklady Akad. Nauk SSSR, 49, 4 (1945).

N. K. Ignatovich, "Regional hydrogeologic patterns in connection with the evaluation of conditions for oil accumulation," Sovetskaya Geologiya, 6 (1945).

N. K. Ignatovich, "Regional subdivisions and the evaluation of conditions of oil accumulation in the Paleozoic strata of the Russian platform as based on hydrogeologic data," Izv. Vses. Geol. Fonda, Gosgeolizdat, 1 (1946).

S. I. Il'in, Conditions for the Formation of Oil in Central Asia. The Origin of Oil and Natural Gas [in Russian] (Press of the Central Scientific Research Institute for Mechanization and Labor Organization in the Petroleum Industry, 1947).

S. I. Il'in, Conditions for the Formation of Oil and the Development of Oil Deposits [in Russian] Lithologic Collection II (State Scientific and Technical Press of the Petroleum and Mineral-Fuel Industry, 1948).

V. N. Ipat'ev, Catalytic Reactions at High Temperatures and Pressures [in Russian] (Acad. Sci. USSR Press, 1936).

B. L. Isachenko, Purple Sulfur Bacteria at the Lower Boundary of the Biosphere [in Russian] Collection: Academy of Sciences, USSR to President V. L. Komarov (1939).

D. J. Jankovski and C. E. ZoBell, "Hydrocarbon production by sulfate-reducing bacteria," Jour. Bacteriology, 47, 5 (1944).

K. P. Kalitskii, "Conditions for the occurrence of oil on the island of Cheleken," Tr. Geol. Komit. Nov. Ser., 59 (1910).

K. P. Kalitskii, Source Beds for Oil [in Russian] (State Scientific and Technical Press of Mining, Geological, and Petroleum Literature, 1934).

K. P. Kalitskii, Scientific Bases for Oil Prospecting [in Russian] (State Scientific and Technical Press of the Petroleum and Mineral-Fuel Industry, 1944).

M. A. Kapelyushnikov, T. P. Zhuze, and S. L. Zaks, "The physical state of oil, gas, and water in petroleum strata," Izv. Akad. Nauk SSSR, Otdelenie Technicheskikh Nauk (Department of Technical Sciences), 11 (1952).

N. M. Karpenko, "New oil fields in the Georgian region (Oisungur and Noiberdy)," Novosti Neftyanoi Techniki, Ser. Geol. 4 (1947).

A. A. Kartsev, "Causes of the systematic distribution of petroleum properties in the fields of the Apsheron Peninsula," Neftyanoe Khozyaistvo, 9, Gostoptekhizdat (1951).

A. A. Kartsev, Z. A. Tabasaranskii, M. I. Subbota, and G. A. Mogilevskii, Geochemical Methods of Prospecting and Exploring for Oil and Gas Fields [in Russian) (State Scientific and Technical Press of the Petroleum and Mineral-Fuel Industry, 1954).

S. M. Katchenkov, "Origin of the elements in petroleum ash," Doklady Akad. Nauk SSSR, 76, 4 (1951).

K. Keil'gak, Ground Water [in Russian] (United Scientific and Technical Press, 1935).

A. M. Kekukh, " The biological basis for the recovery of oleoresin," Priroda, 11 (1954).

G. A. Khel'kvist, Zoned Oil Deposits and Methods of Prospecting for Them [in Russian] (State Scientific and Technical Press of the Petroleum and Mineral-Fuel Industry, 1944).

G. A. Khel'kvist, The Geologic Structure of Zoned Oil Deposits [in Russian] (State Scientific and Technical Press of the Petroleum and Mineral-Fuel Industry, 1946).

L. V. Khmelevskaya, "Hydrogen from depth and its role in the origin of petroleum," Izv. Akad. Nauk SSSR, Ser. Geol., 4 (1947).

N. G. Kholodnyi, "Gases in soils and their biological significance," Priroda, 3 (1953).

Z. A. Kolesnik, "Microflora in the oil fields of the Terek-Dagestan oil district," Tr. VNIGRI, Novaya Seriya, 83, Geological Collection I, Gostoptekhizdat, (1956).

I. D. Korzhnevskii, "Source material and the conditions of formation of oil deposits in the region of Vtori Baku," Neftyanoe Khozyaistvo, 1 (1951).

Yu. A. Kosygin, "A general tectonic classification of salt domes," Neftyanoe Khozyaistvo, 6-7 (1946).

A. L. Kozlov, Problems concerning the Geochemistry of Natural Gas [in Russian] (State Scientific and Technical Press of the Petroleum and Mineral-Fuel Industry, 1950).

A. L. Kozlov, "The gas cycle in the earth's crust and its significance to various branches of geological science," Sovetskaya Geologiya, 83, Geological Collection I, Gostoptekhizdat (1956).

S. A. Kraskovskii, "Thermal conditions in salt domes," Razvedka Nedr, 3 (1937).

S. A. Kraskovskii and Sh. F. Mekhtiev, "Thermal conditions in oil fields," Izv. Akad. Nauk SSSR, 2, Otd. Geol.-Khimich. Nauk i Nefti, 1 (1946).

K. Kreg, "Prospecting for oil," Neftyanoe i Slantsevoe Khozyaistvo [translated from English] (1923).

K. Kreg, "The origin of oil," Neftyanoe i Slantsevoe Khozyaistvo, 1, abstract 1001 (1924).

A. Ya. Krems, "Remarks on the origin of oil and the conditions of its accumulation," Neftyanoe Khozyaistvo, 1 (1947).

A. E. Kriss, and M. N. Lebedeva, "Vertical distribution in abundance and variety of microorganisms in the depth zones of the Black Sea," Doklady Akad. Nauk SSSR, 89, 5 (1953).

P. N. Kropotkin, "Problems concerning the origin of oil," Sovetskaya Geologiya, 47 (1955).

N. A. Kudryavtsev, "The Novogrozny oil district," Izv. Geol-Koma, 43, 2 (1924).

N. A. Kudryavtsev, "Oil fields in the Georgian SSR," Transactions of the 17th International Geological Congress, 4 (1940).

N. A. Kudryavtsev, "Against an organic origin of oil," Neftyanoe Khozyaistvo, 9 (1951).

V. M. Kukanov, "Systematic hydrogeologic relations as factors in setting up studies of deep oil structures," Sovetskaya Geologiya, 4 (1945).

T. A. Kukharenko, "Humic acids in fossil coals," Priroda, 5 (1953).

M. M. Kusakov, P. A. Rebinder, and K. E. Zinchenko, "Surface phenomena in the filtration of oil," Doklady Akad. Nauk SSSR, Novaya Seriya, 28, 5 (1940).

S. I. Kuznetsov, "A study of possible Recent formation of methane and gas and oil facies in the region of Saratov and Buguruslan," Mikrobiologiya, 19, 3 (1950).

N. T. Lindtrop, "Waters from drill holes in the Novogrozny region," Neftyanoe i Slantsevoe Khozyaistvo, 6 (1925).

N. T. Lindtrop, "Characteristics of gushers in the Grozny region," Neftyanoe Khozyaistvo, 9, 10 (1928).

N. T. Lindtrop, "The effect of developing Grozny oil deposits at natural springs," Neftyanoe Khozyaistvo, 8 (1946).

N. T. Lindtrop, The Role of Water in the Formation and Destruction of Oil Deposits [in Russian] Collection in memory of Academician I. M. Gubkin (Acad. Sci. USSR Press, 1951).

V. D. Lomtadze, "Conditions for expressing water and oil from clays," Zapiski Lenin. Gor. In-ta, 25, 2 (1951).

S. P. Maksimov, "The formation of oil deposits in Carboniferous and Devonian rocks at Samarskaya Luka," Neftyanoe Khozyaistvo, 10 (1954).

G. A. Maksimovich, "Character of the oil distribution in the Starogrozny region," Groznenskii Neftyanik, 5-6 (1932).

M. V. Mal'tsev, "Origin of the Tuimazy oils," Neftyanoe Khozyaistvo, 11 (1946).

K. L. Malyarov, The Chemical Composition of Water from Drill Holes in the Grozny Region [in Russian] (Scientific and Technical Press of the Administration of the Supreme Council of the National Economy, 1929).

K. L. Malyarov, "Waters at oil fields," Trudy VNIGRI, Series 10, 46 (1934).

V. T. Malyshek, "Geochemical features of commercial deposits of oil," Transactions of the 17th International Geological Congress, 4 (1940).

V. T. Malyshek, N. D. Sakhnovskaya, and I. Lazarev, Change in the Chemical Composition of Water and Oil in Productive Beds with Increasing Stratigraphic Depth [in Russian] (All-Union Scientific, Engineering, and Technical Society Scientific-Research Papers on Oil, 1, Geology, State Scientific and Technical Press of the Petroleum and Mineral-Fuel Industry, 1943).

Sh. F. Mekhtiev, "Temperature changes in the oil wells at the Ordzhonikidze oil district," Izv. Akad. Nauk Azer. SSR, 11 (1944).

Sh. F. Mekhtiev, "Geothermal observations in deep oil wells," Akad. Nauk Azer. SSR, 2 (1945).

V. S. Melik-Pashaev, "Peculiarities in the chemical composition of formational gases at the oil fields on the Apsheron Peninsula," ANKh 10 (1948).

D. I. Mendeleev, "The origin of oil," ZhRKhO (Journal of the Russian Chemical Society) 9, 1 (1877).

D. I. Mendeleev, Hypothesis Concerning the Origin of Oil [in Russian] Selected Papers, 10 (Acad. Sci. USSR Press, 1949).

M. A. Messineva, New Data Concerning Biochemical Factors in the Transformation of Organic Material During the Formation of Oil [in Russian] Collection: The Origin of Oil and Natural Gas (Information Office for

Technology and Economics. Press of the Central Scientific Research Institute for Mechanization and Labor Organization in the Petroleum Industry, 1947).

M. A. Messineva, Modern Views on the Origin of Oil. The Origin of Oil and Natural Gas [in Russian] (Information Office for Technology and Economics. Press of the Central Scientific Research Institute for Mechanization and Labor Organization in the Petroleum Industry, 1947).

M. F. Mirchink, The Oil Fields of Azerbaidzhan [in Russian] (Azerbaidzhan State United Scientific Press, 1939).

M. F. Mirchink, "The classification of oil fields in Azerbaidzhan according to structural features," ANKh, 5 (1940).

M. F. Mirchink, Stratigraphic Oil Deposits [in Russian](Azerbaidzhan State Technical Press, 1943).

M. F. Mirchink, and A. A. Bakirov, "The geotectonic development of the Russian platform in connection with a study of its oil potential," Neftyanoe Khozyaistvo, 1 (1951).

M. F. Mirchink, A. A. Bakirov, B. F. D'yakov, and D. V. Zhabrev, Editors, The Origin of Oil [in Russian] Collection of Papers (State Scientific and Technical Press of the Petroleum and Mineral-Fuel Industry, 1955).

S. I. Mironov, "The distribution of active oil deposits according to age," Obzor Miner. Resursov, 28, Geol. Kom. (1925).

S. I. Mironov, "The problem of oil formation and the means of solving it," Izv. Acad. Nauk SSSR, Geologiya Seriya, 2 (1952).

S. I. Mironov, "The chief problems discussed in current science concerning the origin of oil," Sovetskaya Geologiya, 47 (1955).

I. V. Mushketov, Physical Geology [in Russian] 2 (St. Petersburg, 1903).

S. S. Nametkin, The Chemistry of Petroleum [in Russian] (Acad. Sci. USSR Press, 1955).

V. M. Nikolaev, The Pattern of Oil Fields in the Terek-Sunzha Oil District and Its Distinctive Features [in Russian] (Grozny Oblast Press, 1945).

V. V. Nikol'skii and B. I. Mineev, "Sapropel and its uses," Priroda, 7 (1953).

S. N. Obryadchikov, Temperature Conditions for the Formation of Oil in Nature [in Russian] Collection: The Origin of Oil and Natural Gas (Information Office for Technology and Economics. Press of the Central Scientific Research Institute for Mechanization and Labor Organization in the Petroleum Industry, 1947).

A. F. Opalev, "The significance of the hydrogeologic factor in evaluating the structure of the Ural-Volga oil province," Sovetskaya Geologiya, 13 (1947).

N. A. Orlov and V. A. Uspenskii, New Information on the Geochemistry of Caustobioliths [in Russian] The Chemistry of Solid Fuels, 8 (1934).

V. B. Porfir'ev, The Problem of Oil Formation in the Light of Current Data [in Russian] (State Scientific and Technical Press of the Petroleum and Mineral-Fuel Industry, 1941).

V. B. Porfir'ev and I. V. Grinberg, "Geochemical basis for the origin of oil," Trudy L'vovskogo Geol. Ob-va, Ser. Geol. 1 (1948).

I. Ya. Postovskii, "New information on the geochemistry of oil," Priroda, 10 (1934).

G. Poton'e, "The formation of coal and oil," Gor. Zhurn. 4, 11 (1906).

G. Poton'e, Sapropelites [in Russian] (United Scientific and Technical Press, 1923).

G. Poton'e, "Critique of current views on the origin of oil," Neftyanoe Khozyaistvo, 10 (1926).

L. V. Pustovalov, Petrography of the Sedimentary Rocks [in Russian] (State Scientific and Technical Press of the Petroleum and Mineral-Fuel Industry, 1941).

O. A. Radchenko, Geochemical Investigations of Mineral Tars in Western Turkmenia [in Russian] The Chemistry of Solid Fuels, 1 (1937).

O. A. Radchenko, Current Views Concerning the Origin of Oil and the Processes of Its Transformation and Destruction [in Russian] Collection: In Memory of Academician I. M. Gubkin (Acad. Sci. USSR Press, 1951).

O. A. Radchenko and L. S. Sheshina, "The origin of porphyrins in oils," Doklady Akad. Nauk SSSR, 105, 6 (1955).

A. S. Razumov, "A direct method for counting bacteria in water. A comparison of this method with the method of Koch," Mikrobiologiya, 1, 2 (1932).

A. B. Ronov, "Hydrogeologic conditions of stability in the gas and oil fields of the Volga region," Doklady Akad. Nauk SSSR, Nov. Ser., 49, 3 (1953).

A. Sakhanov and N. Luchinskii, "Waters in drill holes in the Grozny region," Neftyanoe i Slantsevoe Khozyaistvo, 2 (1924).

B. M. Sarkisyan, The Relationship Between Quality of Oil and the Geologic Environment [in Russian] (Azerbaidzhan Petroleum Press, 1947).

B. M. Sarkisyan, "The formation of oil deposits in the productive sequence of the Apsheron Peninsula," ANKh, 6 (1947).

V. P Savchenko, Natural Gases [in Russian] Collection 9 (United Scientific and Technical Press, 1935).

V. P. Savchenko, "Questions on the formation of oil and gas deposits," Neftyanoe Khozyaistvo, 5 (1952).

L. S. Selivanov, "Geochemistry and biochemistry of disseminated bromine," Trudy Biogeokhim, Laboratorii Acad. Nauk SSSR, 7 (1944).

V. A. Sel'skii, The Migration and Origin of Oil [in Russian] (United Scientific and Technical Press, 1935).

V. A. Sel'skii, Salt Domes and Their Connection with Petroleum [in Russian] (State Scientific and Technical Press, 1936).

A. N. Semikhatov, "Hydrogeologic cycles," Doklady Akad. Nauk SSSR, 56, 6 (1947).

N. T. Shabarova, "Organic acids in Recent marine sediments," Neftyanoe Khozyaistvo, 12 (1953).

N. T. Shabarova, "Oil-forming processes," Priroda, 6 (1954).

A. Shaiderov, "Geothermal observations in the Novogrozny region," Aserb. Neft. Khoz., 4 (1929).

M. Kh. Shakhnazarov, Theory and Practice in the Development of Densely Spaced Oil Fields [in Russian] (Azerbaidzhan Petroleum Press, 1944).

N. S. Shat-skii, Neocatastrophism: The Problem of Orogenic Phases and the Process of Folding [in Russian] (Problems of Soviet Geology 7, 1937).

N. S. Shat-skii, "Orogenic phases of folding," Transactions of the 17th session of the International Geological Congress, 2, GONTI (1939).

N. S. Shat-skii, "The duration of folding movements and the phases of folding," Izv. Akad. Nauk SSSR, Ser. Geol., 1 (1951).

V. N. Shchelkachev and M. A. Gusein-Zade, "The effect of permeability in roof and floor strata on the movement of liquids in them," Neftyanoe Khozyaistvo, 12 (1953).

A. V. Shcherbakov, "Hydrogeologic investigations of subsurface iodine-bromine waters on the island of Cheleken," Izv. VGF, 1 (1945).

A. V. Shcherbakov, "Investigations on waters and brines from drill holes in the Ural-Emba region for purposes of evaluating their possible use in the iodine-bromine industry," Izv. VGF, 1 (1945).

A. V. Shcherbakov, "The nature of the distribution of paraffin and nonparaffin oils in western Turkmenia," Razvedka Nedr, 3 (1946).

L. D. Shturm, "Data on microbiological investigations on the oil fields of Vtoroi Baku," Trudy Instituta Nefti, 1 (1950).

A. I. Silin-Bekchurin, "The formation of the mineral waters of Bashkiria," Doklady Akad. Nauk SSSR, 51, 1 (1946).

P. V. Smit, Investigations on the Origin of Oil [in Russian] (State Scientific and Technical Press of the Petroleum and Mineral-Fuel Industry, 1956).

V. A. Sokolov, "Radioactivity and the origin of oil," Transactions of the 17th session of the International Geological Congress (1937), 4, Gostoptekhizdat (1940).

V. A. Sokolov, Essays on the Origin of Oil [in Russian] (State United Scientific and Technical Press, 1948).

V. D. Sokolov, The Cosmic Origin of Oil [in Russian] (1913).

Yu. I. Sorokin, "New ways of isolating sulfate-reducing bacteria," Trudy Inst. Mikrobiologii, Institute Mikrobiologii Akad. Nauk SSSR, 2 (1952).

G L. Stadnikov, The Origin of Coal and Oil [in Russian] (Acad. Sci. USSR Press, 1937).

P. I. Stepanov and S. I. Mironov, The Geology of Caustobiolith Deposits [in Russian] (United Scientific and Technical Press, 1937).

N. M. Strakhov, "The significance of hydrogen-sulfide basins as zones for accumulation of bituminous strata and of source beds for petroleum," Izv. Akad. Nauk SSSR, Ser. Geol., 5 (1937).

N. M. Strakhov, "Climating zoning during late Paleozoic time in northwestern Eurasia," Sovetskaya Geologiya, 6 (1945).

N. M. Strakhov, Principles of Historical Geology [in Russian] (State Press of Geological Literature, 1948).

N. I. Strizhov and I. E. Khodanovich, The Recovery of Gas [in Russian] (State Scientific and Technical Press of the Petroleum and Mineral-Fuel Industry, 1946).

M. I. Subbota, "On the problem of the origin of oil," Neftyanoe Khozyaistvo, 11 (1952).

E. I. Sukhankin, "Gases dissolved in Devonian water," Bashkirskaya Neft', 3 (1947).

G. M. Sukharev, The Waters at Oil and Gas Fields in Eastern Ciscaucasia [in Russian] (Grozny Oblast Press, 1947).

G. M. Sukharev, "Hydrogeologic conditions of forming oil and gas deposits in the Terek-Dagestan oil province," Neftyanoe Khozyaistvo, 10 (1947).

G. M Sukharev, Geothermal Peculiarities of the Terek-Dagestan Oil-Gas Province [in Russian] (State Scientific and Technical Press of the Petroleum and Mineral-Fuel Industry, 1948).

G. M. Sukharev, The Hydrogeological Conditions for Forming Oil and Gas Deposits in the Terek-Dagestan Oil and Gas Province [in Russian] (Grozny Oblast Press, 1948).

G. M. Sukharev, "The sequence of strata in the Tashkala oil field," Trudy Groznen. Neft. In-ta, Collection 11 (1948).

G. M. Sukharev, Hydrogeology of the Mesozoic and Tertiary Strata in the Terek-Dagestan Oil and Gas Region and in the Lower Volga Region [in Russian] (State Scientific and Technical Press of the Petroleum and Mineral-Fuel Industry, 1954).

V. A. Sulin, Oil-Field Waters in the USSR [in Russian] (State United Scientific and Technical Press, 1936).

V. A. Sulin, Waters at Oil Fields in the System of Natural Waters [in Russian] (State Scientific and Technical Press of the Petroleum and Mineral-Fuel Industry, 1946).

V. A. Sulin, The Hydrogeology of Oil Fields [in Russian] (State Scientific and Technical Press of the Petroleum and Mineral-Fuel Industry, 1948).

Z. A. Tabasaranskii, "Conditions of forming oil and gas deposits in the middle and upper Paleogene strata of the Il'sko-Kholmskii region," Neftyanoe Khozyaistvo, 4 (1954).

N. V. Tageeva, "The origin of waters at oil deposits," Neftyanoe Khozyaistvo, 7 (1934).

V. O. Tauson, "The decomposition of hydrocarbons by microorganisms," Priroda, 6 (1934).

G. I. Teodorovich, "Geochemical and other conditions favorable for the development of petroleum bitumens," Neftyanoe Khozyaistvo, 12 (1952).

M. P. Tolstoi, "The oil potential in Middle Devonian strata in the region of the Moscow basin," Neftyanoe Khozyaistvo, 7 (1945).

A. Traibs, "Chlorophyll and derived hematin in asphaltic petroleum rocks, in waxes, and in asphalts," [in German] Libichs Annalen der Chemie (1934).

A. Traibs, "Plant substances as the source material of petroleum," [in German] Brennschtoff Geologie (1935).

P. D. Trask, "Source material for oil," Transactions of the 17th session of the International Geological Congress, 4, Gostoptekhizdat (1940).

P. D. Trask and H. W. Patnode, "The origin of oil," Am. Assoc. Petroleum Geologists Bulletin (1942).

B. A. Trofimov, "Life during geologic epochs," Priroda, 6 (1954).

Trudy VNIGRI, Nov. Seriya, 83, Geological Collection, 1, Gostoptekhizdat (1955).

A. Uiks, "Some experiments to test the hydraulic theory of migration and accumulation of oil by descending circulating waters," Neftyanoe Khozyaistvo, 8 (1926).

A. S. Uklonskii, "The paragenesis of sulfur and petroleum," Transactions of the 17th session of the International Geological Congress, 4, Akad. Nauk Uzbek SSR (1940).

A. V. Ul'yanov, Source-Bed Facies in the Northwestern Caucasus [in Russian] The Origin of Oil and Natural Gas (Information Office for Technology and Economics. Press of the Central Scientific Research Institute for Mechanization and Labor Organization in the Petroleum Industry, 1947).

A. V. Ul'yanov and G. A. Khal'kvist, The Geology of Oil and Gas Fields [in Russian] (State Scientific and Technical Press of the Petroleum and Mineral-Fuel Industry, 1955).

V. A. Uspenskii, A. S. Chernysheva, and Yu. A. Mandrykina, "Discoveries of disseminated hydrocarbons in various sedimentary rocks," Izv. Akad. Nauk SSSR, Ser. Geol., 5 (1949).

V. A. Uspenskii and O. A. Radchenko, The Origin of Types of Oil [in Russian] (Leningrad Branch of the State Scientific and Technical Press of the Petroleum and Mineral-Fuel Industry, 1947).

V. A. Uspenskii and O. A. Radchenko, "The origin of oil," Neftyanoe Khozyaistvo, 8 (1954).

F. M. Van Tuyl, B. H. Parker, and W. W. Skeeters, "The migration and accumulation of petroleum and natural gas," [Russian translation] (Foreign Literature Press, 1948).

V. V. Veber, "Discussions on petroliferous facies in the Produktivnyy (Productive) sequence," ANKh, 9 (1946).

V. V. Veber, Petroliferous Facies and Their Role in the Formation of Oil Deposits [in Russian] (Leningrad Branch of the State Scientific and Technical Press of the Petroleum and Mineral-Fuel Industry, 1947).

V. V. Veber, The Problem of Oil Formation in the Light of Paleogeographic Data of Oil-Producing Basins. Collection "Origin of Oil and Natural Gas" [in Russian] (Information Office for Technology and Economics. Press of the Central Scientific Research Institute for Mechanization and Labor Organization in the Petroleum Industry, 1947).

V. V. Veber and N. T. Shabarova, "Alteration of organic material in Recent marine sediments during their transformation," Trudy. Mosk. Filiala VNIGRI, 2, Gostoptekhizdat (1953).

V. I. Vernadskii, "The question of radioactivity in waters in oil wells," Doklady Akad. Nauk SSSR, 15 (1930).

B. N. Viktorov, "The effect of centrifugal force of the earth's rotation on the formation of oil and gas deposits," Neftyanoe Khozyaistvo, 2 (1954).

A. P. Vinogradov, The Geochemistry of Rare and Disseminated Chemical Elements in Soils [in Russian] (Acad. Sci. USSR Press, 1950).

S. A. Waksman, "Distribution of organic material on sea floors and the chemical nature and origin of marine humus," Soil Science, 5, 36 (1933).

S. L. Zaks, "Buried water and its significance in petroleum production," Neftyanoe Khozyaistvo, 4 (1947).

S. L. Zaks, "The effect of rocks and their associated water on the magnitude of the pressure in the transition of the system oil-gas to the single-phase gaseous state," Doklady. Akad. Nauk SSSR, 86, 5 (1952).

N. D. Zelinskii, Balkhash Sapropelite and the Hypothesis of an Organic Origin of Oil [in Russian] Selected Works, 1, (Acad. Sci. USSR Press, 1941).

N. D. Zelinskii, Oleic, Palmitic, and Stearic Acids as Source Material of Petroleum [in Russian] Selected Works, 1 (Acad. Sci. USSR Press, 1941).

N. D. Zelinskii, The Chemical Nature of the Hydrocarbons Derived from the Decomposition of Elaterite [in Russian] Selected Works, 1 (Acad. Sci. USSR Press, 1941).

N. D. Zelinskii, Cholesterol as Source Material for Oil, Collections 1 and 2 [in Russian] Selected Works, 1 (Acad. Sci. USSR Press, 1941).

N. D. Zelinskii, Some Notes on the Origin of Oil. Collection: Papers in Memory of Academician V. I. Vernadskii [in Russian] (Acad. Sci. USSR Press, 1946).

D. V. Zhabrev, "Changes in the composition of water about oil fields as a whole and in specific horizons," Trudy Geol. In-ta im. Gubkina, 19, Azer. Filial Akad. Nauk SSSR (1939).

C. E. ZoBell, "Bacterial life at the bottom of the Phillippine trench," Science, 115, 2993 (1952).

C. E. ZoBell, and F. H. Johnson, "The influence of hydrostatic pressure on the growth and viability of terrestrial and marine bacteria," Jour. Bacteriology, 57, 2 (1949).